Use R!

Series Editors:
Robert Gentleman   Kurt Hornik   Giovanni Parmigiani

Use R!

Albert: Bayesian Computation with R
Cook/Swayne: Interactive and Dynamic Graphics for Data Analysis
Paradis: Analysis of Phylogenetics and Evolution with R
Pfaff: Analysis of Integrated and Cointegrated Time Series with R

Dianne Cook   Deborah F. Swayne

Interactive and Dynamic Graphics for Data Analysis

With R and GGobi

With Contributions by Andreas Buja, Duncan Temple Lang,
Heike Hofmann, Hadley Wickham, and Michael Lawrence

Dianne Cook
Department of Statistics
Iowa State University
325 Snedecor Hall
Ames, IA 50011-1210
dicook@iastate.edu

Deborah F. Swayne
AT & T Labs - Research
Shannon Laboratory
180 Park Avenue
Florham Park, NJ 07932-1049
dfs@research.att.com

Series Editors:
Robert Gentleman
Program in Computational Biology
Division of Public Health Sciences
Fred Hutchinson Cancer Research Center
1100 Fairview Ave. N, M2-B876
Seattle, Washington 981029-1024
USA

Kurt Hornik
Department für Statistik und Mathematik
Wirtschaftsuniversität Wien Augasse 2-6
A-1090 Wien
Austria

Giovanni Parmigiani
The Sidney Kimmel Comprehensive Cancer Center at Johns Hopkins University
550 North Broadway
Baltimore, MD 21205-2011
USA

ISBN 978-0-387-71761-6   e-ISBN 978-0-387-71762-3

Library of Congress Control Number: 2007925720

© 2007 Springer Science+Business Media, LLC
All rights reserved. This work may not be translated or copied in whole or in part without the written permission of the publisher (Springer Science+Business Media, LLC, 233 Spring Street, New York, NY 10013, USA), except for brief excerpts in connection with reviews or scholarly analysis. Use in connection with any form of information storage and retrieval, electronic adaptation, computer software, or by similar or dissimilar methodology now known or hereafter developed is forbidden.
The use in this publication of trade names, trademarks, service marks, and similar terms, even if they are not identified as such, is not to be taken as an expression of opinion as to whether or not they are subject to proprietary rights.

Cover illustration: Brushing a cluster that was found by using the grand tour to explore two-dimensional projections of five-dimensional data on Australian crabs.

HTML HyperText Markup Language; claimed as a trademark or generic term by MIT, ERCIM, and/or Keio on behalf of the W3C. Data Desk is a registered trademark of Data Description, Inc. Spotfire is a registered trademark of Spotfire, Inc. JMP is a registered trademark of SAS Institute Inc. Linux is a registered trademark of Linus Torvalds, the original author of the Linux kernel. Mac OS X - Operating System software - is a registered trademark of Apple Computer, Inc. S-PLUS is a registered trademark of Insightful Corporation. Microsoft® is a registered trademark of Microsoft Corporation. Windows is a registered trademarks of Microsoft Corporation. XML - Extensible Markup Language; Language by W3C - claimed as a trademark or generic term by MIT, ERCIM, and/or Keio on behalf of the W3C. Facebook is a registered trademark of Facebook Inc. MySpace is a Registered Trademark of News Corp. SYSTAT is a registered trademark of Systat Software, Inc. SPSS is a registered trademark of SPSS Inc.

Printed on acid-free paper.

9 8 7 6 5 4 3 2 1

springer.com

Preface

This book is about using interactive and dynamic plots on a computer screen as part of data exploration and modeling, both alone and as a partner with static graphics and non-graphical computational methods. The area of interactive and dynamic data visualization emerged within statistics as part of research on exploratory data analysis in the late 1960s, and it remains an active subject of research today, as its use in practice continues to grow. It now makes substantial contributions within computer science as well, as part of the growing fields of information visualization and data mining, especially visual data mining.

The material in this book includes:

- An introduction to data visualization, explaining how it differs from other types of visualization.
- A description of our toolbox of interactive and dynamic graphical methods.
- An approach for exploring missing values in data.
- An explanation of the use of these tools in cluster analysis and supervised classification.
- An overview of additional material available on the web.
- A description of the data used in the analyses and exercises.

The book's examples use the software R and GGobi. R (Ihaka & Gentleman 1996, R Development Core Team 2006) is a free software environment for statistical computing and graphics; it is most often used from the command line, provides a wide variety of statistical methods, and includes high–quality static graphics. R arose in the Statistics Department of the University of Auckland and is now developed and maintained by a global collaborative effort. It began as a re-implementation of the S language and statistical computing environment (Becker & Chambers 1984) first developed at Bell Laboratories before the breakup of AT&T.

GGobi (Swayne, Temple Lang, Buja & Cook 2003) is free software for interactive and dynamic graphics; it can be operated using a command-line interface or from a graphical user interface (GUI). When GGobi is used as a stand-alone tool, only the GUI is used; when it is used with R, via the rggobi (Temple Lang, Swayne, Wickham & Lawrence 2006) package, a command-line interface is used along with the GUI. GGobi is a descendant of two earlier programs: XGobi (Swayne, Cook & Buja 1992, Buja, Cook & Swayne 1996) and, before that, Dataviewer (Buja, Hurley & McDonald 1986, Hurley 1987). Many of the examples that follow might be reproduced with other software such as S-PLUS®, SAS JMP®, DataDesk®, Mondrian, MANET, and Spotfire®. However, GGobi is unique because it offers tours (rotations of data in higher than 3D), complex linking between plots using categorical variables, and the tight connection with R.

Web resources

The web site which accompanies the book contains sample datasets and R code, movies demonstrating the interactive and dynamic graphic methods, and additional chapters. It can be reached through the GGobi web site:
http://www.ggobi.org
The R software is available from:
http://www.R-project.org
Both web sites include source code as well as binaries for various operating systems (Linux®, Windows®, Mac OS X®) and allow users to sign up for mailing lists and browse mailing list archives. The R web site offers a wealth of documentation, including an introduction to R and a partially annotated list of books offering more instruction. [Widely read books include Dalgaard (2002), Venables & Ripley (2002), and Maindonald & Braun (2003).] The GGobi web site includes an introductory tutorial, a list of papers, and several movies.

How to use this book

The language in the book is aimed at later year undergraduates, beginning graduate students, and graduate students in any discipline needing to analyze their own multivariate data. It is suitable reading for an industry statistician, engineer, bioinformaticist, or computer scientist with some knowledge of basic data analysis and a need to analyze high-dimensional data. It also may be useful for a mathematician who wants to visualize high-dimensional structures.

The end of each chapter contains exercises to help practice the methods discussed in the chapter. The book may be used as a text in a class on statistical graphics, exploratory data analysis, visual data mining, or information visualization. It might also be used as an adjunct text in a course on multivariate data analysis or data mining.

This book has been closely tied to a particular software implementation so that you can actively use the methods as you read about them, to learn and experiment with interactive and dynamic graphics. The plots and written explanations in the book are no substitute for personal experience. We strongly urge the reader to go through this book sitting near a computer with

GGobi, R, and `rggobi` installed, following along with the examples. If you do not wish to install the software, then the next best choice is to watch the accompanying movies demonstrating the examples in the text.

If you have not used GGobi before, then visit the web site, watch the movies, download the manual, and work through the tutorial; the same advice applies for those unfamiliar with R: Visit the R web site and learn the basics.

As you read the book, try things out for yourself. Take your time, and have fun!

Acknowledgments

This work started at Bellcore (now Telcordia), where Jon Kettering, Ram Gnanadesikan, and Diane Duffy carefully nurtured graphics research in the 1990s. It has continued with the encouragement of our respective employers at AT&T (with support from Daryl Pregibon and Chris Volinsky), at Lucent Bell Laboratories (with support from Diane Lambert), and at Iowa State University (with support from chairs Dean Isaacson and Ken Koehler). Werner Stuetzle provided the initial encouragement to write this book. A large part of the book was drafted while Dianne Cook was partially supported for a semester by Macquarie University, Sydney.

The authors would like to thank the peer reviewers recruited by the publisher for their many useful comments.

Proceeds from the sale of each book will be directed to the GGobi Foundation, to nurture graphics research and development.

Dianne Cook
Iowa State University

Deborah F. Swayne
AT&T Labs – Research

July 2007

Contents

Preface V

Technical Notes XIII

List of Figures XV

1 Introduction 1
   1.1 Data visualization: beyond the third dimension 1
   1.2 Statistical data visualization: goals and history 3
   1.3 Getting down to data 4
   1.4 Getting real: process and caveats 8
   1.5 Interactive investigation 15

2 The Toolbox 17
   2.1 Introduction 17
   2.2 Plot types 19
       2.2.1 Univariate plots 19
       2.2.2 Bivariate plots 21
       2.2.3 Multivariate plots 24
       2.2.4 Plot arrangement 34
   2.3 Plot manipulation and enhancement 35
       2.3.1 Brushing 35
       2.3.2 Identification 41
       2.3.3 Scaling 41
       2.3.4 Subset selection 42
       2.3.5 Line segments 43
       2.3.6 Interactive drawing 43
       2.3.7 Dragging points 43
   2.4 Tools available elsewhere 44
   2.5 Recap 45
   Exercises 45

3 Missing Values ... 47
- 3.1 Background ... 48
- 3.2 Exploring missingness ... 49
  - 3.2.1 Shadow matrix ... 49
  - 3.2.2 Getting started: missings in the "margins" ... 52
  - 3.2.3 A limitation ... 53
  - 3.2.4 Tracking missings using the shadow matrix ... 55
- 3.3 Imputation ... 55
  - 3.3.1 Mean values ... 56
  - 3.3.2 Random values ... 56
  - 3.3.3 Multiple imputation ... 58
- 3.4 Recap ... 61
- Exercises ... 62

4 Supervised Classification ... 63
- 4.1 Background ... 64
  - 4.1.1 Classical multivariate statistics ... 65
  - 4.1.2 Data mining ... 66
  - 4.1.3 Studying the fit ... 69
- 4.2 Purely graphics: getting a picture of the class structure ... 70
  - 4.2.1 Overview of Italian Olive Oils ... 70
  - 4.2.2 Building classifiers to predict region ... 71
  - 4.2.3 Separating the oils by area within each region ... 73
  - 4.2.4 Taking stock ... 77
- 4.3 Numerical methods ... 77
  - 4.3.1 Linear discriminant analysis ... 77
  - 4.3.2 Trees ... 81
  - 4.3.3 Random forests ... 83
  - 4.3.4 Neural networks ... 88
  - 4.3.5 Support vector machine ... 92
  - 4.3.6 Examining boundaries ... 97
- 4.4 Recap ... 99
- Exercises ... 99

5 Cluster Analysis ... 103
- 5.1 Background ... 105
- 5.2 Purely graphics ... 107
- 5.3 Numerical methods ... 111
  - 5.3.1 Hierarchical algorithms ... 111
  - 5.3.2 Model-based clustering ... 113
  - 5.3.3 Self-organizing maps ... 119
  - 5.3.4 Comparing methods ... 122
- 5.4 Characterizing clusters ... 125
- 5.5 Recap ... 126
- Exercises ... 127

6 Miscellaneous Topics ... 129
- 6.1 Inference ... 129
- 6.2 Longitudinal data ... 134
- 6.3 Network data ... 139
- 6.4 Multidimensional scaling ... 145
- Exercises ... 151

7 Datasets ... 153
- 7.1 Tips ... 153
- 7.2 Australian Crabs ... 154
- 7.3 Italian Olive Oils ... 155
- 7.4 Flea Beetles ... 157
- 7.5 PRIM7 ... 157
- 7.6 Tropical Atmosphere-Ocean Array (TAO) ... 159
- 7.7 Primary Biliary Cirrhosis (PBC) ... 161
- 7.8 Spam ... 162
- 7.9 Wages ... 164
- 7.10 Rat Gene Expression ... 166
- 7.11 Arabidopsis Gene Expression ... 168
- 7.12 Music ... 171
- 7.13 Cluster Challenge ... 172
- 7.14 Adjacent Transposition Graph ... 172
- 7.15 Florentine Families ... 173
- 7.16 Morse Code Confusion Rates ... 174
- 7.17 Personal Social Network ... 175

References ... 177

Index ... 185

Technical Notes

R code

The R code in this book, denoted by `typewriter font`, and the more extensive code on the web site, has been tested on version 2.4.0 of R, version 2.1.5 of GGobi, and version 2.1.5 of `rggobi`. Updates will be available on the web site as they are needed.

Figures

The figures in this book were produced in a variety of ways, and the files and code to reproduce them are all available on the book's web site. Some were produced directly in R. Some were produced using both GGobi and R, and the process of converting GGobi views into publication graphics deserves an explanation.

When we arrive at a GGobi view we want to include in a paper or book, we use the `Save Display Description` item on GGobi's `Tools` menu to generate a file containing an S language description of the display. We read the file into R using the R package `DescribeDisplay` (Wickham 2006*b*), like this:

```
> library(DescribeDisplay)
> d <- dd_load("fig.R")
```

We create the publication-quality graphic using either that package's plot method or another R package, `ggplot` (Wickham 2006*c*), like this:

```
> plot(d)
```

or

```
> p <- ggplot(d)
> print(p)
```

Figure 0.1 illustrates the differences with a trio of representations of the same bivariate scatterplot. The picture at left is a screen dump of a GGobi display. Such images are not usually satisfactory for publication for several

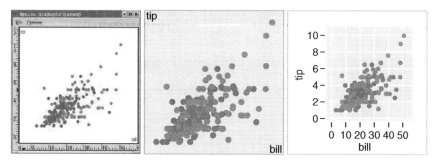

Fig. 0.1. Sample plots produced from GGobi in different ways: **(left)** a simple screen dump; **(middle)** a plot produced using the plot method of the R package DescribeDisplay; **(right)** a plot made using the R package ggplot.

reasons, the most obvious of which is the lack of resolution. The second picture was produced using DescribeDisplay's plot method, which reproduces the plotting region of the view with pretty good fidelity. We used this method to produce most of the one–dimensional and two-dimensional tour pictures in this book. The third picture was produced using ggplot, which adds axis ticks, labels and grid lines. We used it to produce nearly all the bivariate scatterplots of GGobi views in this book.

List of Figures

0.1 Sample plots produced from GGobi XIV

1.1 Histograms of tip with differing bin width 6
1.2 Scatterplot of tip vs. bill 8
1.3 Scatterplot of tip vs. bill conditioned by sex and smoker 9
1.4 Factors affecting tipping behavior 12
1.5 Histogram of tip linked to 2D mosaic plots of sex and smoker .. 16

2.1 Univariate displays 20
2.2 Barchart and spine plot of the Tips data 22
2.3 Bivariate scatterplots for the Blue female Australian Crabs 22
2.4 A mosaic plot of day and sex for the Tips data 23
2.5 Parallel coordinate plot of the Australian Crabs data 24
2.6 Scatterplot matrix of the real-valued variables in Australian Crabs 25
2.7 One-dimensional projections of the Crabs data 27
2.8 Two-dimensional projections of the Crabs data 28
2.9 Two-dimensional projections found by projection pursuit 32
2.10 One-dimensional projections found by projection pursuit 33
2.11 The relationship between a 2D tour and the biplot............ 35
2.12 A dynamic query posed using linked brushing 36
2.13 Categorical brushing 38
2.14 Linking by subject identifier 39
2.15 Linking between a point in one plot and a line in another, using Arabidopsis Gene Expression 40
2.16 Transient brushing contrasted with persistent painting 40
2.17 Identifying points in a plot using three different labeling styles . 41
2.18 Scaling a plot reveals different aspects of the River Flow data... 42

3.1 Assigning constants to missing values 52
3.2 Tour views with missings set to 10% below minimum 54

3.3	Parallel coordinates plot with missings set to 10% below minimum	54
3.4	Exploring missingness in the TAO data	56
3.5	Comparison of simple, widely used imputation schemes	57
3.6	Conditional random imputation	58
3.7	Two different imputations using simulation from a multivariate normal distribution of all missing values	60
3.8	Tour projection of the data after multiple imputation	61
4.1	Evaluating model assumptions for 2D data	67
4.2	Classification of the data space for the Flea Beetles	68
4.3	Differentiating the oils from the three regions in the Olive Oils data in univariate plots	71
4.4	Separation between the Northern and Sardinian oils	72
4.5	Finding variables for classification using a parallel coordinate plot of Olive Oils	73
4.6	Separating the oils of the Northern region by area	75
4.7	Separating the oils of the Southern region by area	76
4.8	Evaluating model assumptions for the 6D Flea Beetles data	78
4.9	Misclassifications from an LDA classifier of the Olive oils.	79
4.10	Improving on the results of the tree classifier using the manual tour	82
4.11	Examining the results of a forest classifier of Olive Oils	84
4.12	Examining the results of a random forest after classifying the oils of the South	87
4.13	Misclassifications of a feed-forward neural network classifying the oils from the South	91
4.14	Misclassifications of a support vector machine classifying the oils of the South	96
4.15	Using the tour to examine the choice of support vectors	96
4.16	Classification boundaries for different models	98
5.1	Structures in data and their impact on cluster analysis	104
5.2	Clustering the example data	106
5.3	Stages of "spin and brush" on PRIM7	109
5.4	The PRIM7 model summarized	110
5.5	Hierarchical clustering of the particle physics data	112
5.6	Model-based clustering of a reduced set of Australian Crabs	115
5.7	Comparing the Australian Crabs data with results of model-based clustering	118
5.8	Comparison of clustering music tracks using a self-organizing map versus principal components	120
5.9	Unsuccessful SOM fit	122
5.10	Successful SOM fit	123

List of Figures   XVII

5.11 Comparing two five-cluster models of the Music data using
     confusion tables linked to tour plots 125
5.12 Characterizing clusters in a parallel coordinate plot 126

6.1 Comparison of various pairwise relationships with the same
     correlation 130
6.2 Testing the assumption that species labels correspond to
     groups in the data 132
6.3 Testing the assumption that species labels correspond to
     groups in simulated data 133
6.4 Exploring the longitudinal Wages data 135
6.5 Using linked brushing to explore subject profiles 138
6.6 Business and marital relationships among Renaissance
     Florentine families 141
6.7 Highlighting the links involving the greatest number of ties ... 142
6.8 The $n = 3$ adjacency transposition graph 143
6.9 The $n = 4$ adjacency transposition graph in various layouts ... 144
6.10 A grid and its distance matrix 146
6.11 Snapshots from MDS animation 148
6.12 Applying MDS to the Morse Code confusion data 149
6.13 Studying MDS results 150
6.14 3D MDS layouts of the Morse Code 150

1
Introduction

In this technological age, we live in a sea of information. We face the problem of gleaning useful knowledge from masses of words and numbers stored in computers. Fortunately, the computing technology that produces this deluge also gives us some tools to transform heterogeneous information into knowledge. We now rely on computers at every stage of this transformation: structuring and exploring information, developing models, and communicating knowledge.

In this book we teach a methodology that makes visualization central to the process of abstracting knowledge from information. Computers give us great power to represent information in pictures, but even more, they give us the power to interact with these pictures. If these are pictures of data, then interaction gives us the feeling of having our hands on the data itself and helps us to orient ourselves in the sea of information. By generating and manipulating many pictures, we make comparisons among different views of the data, we pose queries about the data and get immediate answers, and we discover large patterns and small features of interest. These are essential facets of data exploration, and they are important for model development and diagnosis.

In this first chapter we sketch the history of computer-aided data visualization and the role of data visualization in the process of data analysis.

1.1 Data visualization: beyond the third dimension

So far we have used the terms "information," "knowledge," and "data" informally. From now on we will use the following distinction: the term *data* refers to information that is structured in some schematic form such as a table or a list, and knowledge is derived from studying data. Data is often but not always quantitative, and it is often derived by processing unstructured information. It always includes some attributes or variables such as the number of hits on web sites, frequencies of words in text samples, weight in pounds, mileage in gallons per mile, income per household in dollars, years of education, acidity on the pH scale, sulfur emissions in tons per year, or scores on standardized tests.

When we visualize data, we are interested in portraying abstract relationships among such variables: for example, the degree to which income increases with education, or whether certain astronomical measurements indicate grouping and therefore hint at new classes of celestial objects. In contrast to this interest in abstract relationships, many other areas of visualization are principally concerned with the display of objects and phenomena in physical three-dimensional (3D) space. Examples are volume visualization (e.g., for the display of human organs in medicine), surface visualization (e.g., for manufacturing cars or animated movies), flow visualization (e.g., for aeronautics or meteorology), and cartography (e.g., for navigation or social studies). In these areas one often strives for physical realism or the display of great detail in space, as in the visual display of a new car design or of a developing hurricane in a meteorological simulation. The data visualization task is obviously different from drawing physical objects.

If data visualization emphasizes abstract variables and their relationships, then the challenge of data visualization is to create pictures that reflect these abstract entities. One approach to drawing abstract variables is to create axes in space and map the variable values to locations on the axes, and then render the axes on a drawing surface. In effect, one codes non-spatial information using spatial attributes: position and distance on a page or computer screen. The goal of data visualization is then not realistic drawing, which is meaningless in this context, but translating abstract relationships to interpretable pictures.

This way of thinking about data visualization, as interpretable spatial representation of abstract data, immediately brings up a limitation: Plotting surfaces such as paper or computer screens are merely two-dimensional (2D). We can extend this limit by simulating a third dimension: The eye can be tricked into seeing 3D virtual space with perspective and motion, but if we want an axis for each variable, that's as far as we can stretch the display dimension.

This limitation to a 3D display space is not a problem if the objects to be represented are three-dimensional, as in most other visualization areas. In data visualization, however, the number of axes required to code variables can be large: Five or ten axes are common, but these days one often encounters dozens and even hundreds. Overcoming the 2D and 3D barriers is a key challenge for data visualization. To meet this challenge, we use powerful computer-aided visualization tools. For example, we can mimic and amplify a strategy familiar from photography: taking pictures from multiple directions so the shape of an object can be understood in its entirety. This is an example of the "multiple views" paradigm, which will be a recurring theme of this book. In our 3D world the paradigm works superbly, because the human eye is adept at inferring the true shape of an object from just a few directional views. Unfortunately, the same is often not true for views of abstract data. The chasm between different views of data, however, can be actively bridged with additional computer technology: Unlike the passive paper medium, computers allow us to manipulate pictures, to pull and push their content in continuous motion like a moving video camera, or to poke at objects in one picture and see them light up in other pictures. Motion links pictures in time; poking links them across space. This book features many illustrations of the power of these linking technologies. The diligent reader may come away "seeing" high-dimensional data spaces!

1.2 Statistical data visualization: goals and history

Visualization has been used for centuries to map our world (cartography) and describe the animal and plant kingdoms (scientific illustration). Data visualization, which is more abstract, emerged more recently. An early innovator was William Playfair, whose extensive charting of economic data in the 1800s (Wainer & Spence 2005a, Wainer & Spence 2005b) contributed to its emergence. The early history of visualization has been richly — and beautifully — documented (Friendly & Denis 2006, Tufte 1983, Tufte 1990, Ford 1992, Wainer 2000).

Today's data visualization has homes in several disciplines, including the natural sciences, engineering, geography, computer science, and statistics. There is a lot of overlap in the functionality of the methods and tools they generate, but some differences in emphasis can be traced to the research contexts in which they were incubated. For example, data visualization in the natural science and engineering communities supports the goal of modeling physical objects and processes, relying on scientific visualization (Hansen & Johnson 2004, Bonneau, Ertl & Nielson 2006). For the geographical community, maps are the starting point, and other data visualization methods are used to expand on the information displayed using cartography (Longley, Maguire, Goodchild & Rhind 2005, Dykes, MacEachren & Kraak 2005). The database research community creates visualization software that grows out of their work in data storage and retrieval; their graphics often summarize the kinds of tables and tabulations that are common results of database queries. The human–computer interface community produces software as part of their research in human perception, human–computer interaction and usability, and their tools are often designed to make the performance of a complex task as straightforward as possible. These two latter fields have been instrumental in developing the field of information visualization (Card, Mackinlay & Schneiderman 1999, Bederson & Schneiderman 2003, Spence 2007).

The statistics community creates visualization systems within the context of data analysis, so the graphics are designed to support and enrich the statistical processes of data exploration, modeling, and inference. As a result, statistical data visualization has some unique features. Statisticians are always concerned with variability in observations and error in measurements, both of which cause uncertainty about conclusions drawn from data. Dealing with this uncertainty is at the heart of classical statistics, and statisticians have developed a huge body of inferential methods that help to quantify uncertainty.

Systems for data analysis included visualization as soon as they began to emerge (Nie, Jenkins, Steinbrenner & Bent 1975, Becker & Chambers 1984, Wilkinson 1984). They could generate a wide variety of plots, either for display on the screen or for printing, and the more flexible systems have always allowed users considerably leeway in plot design. Since these systems predated the general use of the mouse, the keyboard was their only input device, and the displays on the screen were not themselves interactive.

As early as the 1960s, however, researchers in many disciplines were making innovations in computer–human interaction, and statisticians were there, too. The seminal visualization system for exploratory data analysis was PRIM-9, the work of Fisherkeller, Friedman, and Tukey at the Stanford Linear Accelerator Center in 1974. PRIM-9 was the first stab at an interactive tool set for the visual analysis of multivariate data. It was followed by further pioneering systems at the Swiss Federal Institute of Technology (PRIM-ETH), Harvard University (PRIM-H), and Stanford University (ORION), in the late 1970s and early 1980s.

Research picked up in the following few years in many places (Wang 1978, McNeil 1977, Velleman & Velleman 1985, Cleveland & McGill 1988, Buja & Tukey 1991, Tierney 1991, Cleveland 1993, Rao 1993, Carr, Wegman & Luo 1996). The authors themselves were influenced by work at Bell Laboratories, Bellcore, the University of Washington, Rutgers University, the University of Minnesota, MIT, CMU, Batelle Richmond WA, George Mason University, Rice University, York University, Cornell University, Trinity College, and the University of Augsburg, among others. In the past couple of years, books have begun to appear that capture this history and continue to point the way forward (Wilkinson 2005, Young, Valero-Mora & Friendly 2006, Unwin, Theus & Hofmann 2006, Chen, Härdle & Unwin 2007).

1.3 Getting down to data

Here is a very small and seemingly simple dataset we will use to illustrate the use of data graphics. One waiter recorded information about each tip he received over a period of a few months working in one restaurant. He collected several variables:

- tip (i.e., gratuity) in US dollars
- bill (the cost of the meal) in US dollars
- sex of the bill payer
- whether the party included smokers
- day of the week
- time of day
- size of the party

In all he recorded 244 tips. The data was reported in a collection of case studies for business statistics (Bryant & Smith 1995). The primary question suggested by the data is this: *What are the factors that affect tipping behavior?*

This dataset is typical (albeit small): There are seven variables, of which two are numeric (tip, bill), and the others are categorical or otherwise discrete. In answering the question, we are interested in exploring relationships that may involve more than three variables, none of which corresponds to physical space. In this sense the data is high-dimensional and abstract.

We look first at the variable of greatest interest to the waiter: tip. A common graph for looking at a single variable is the histogram, where data values are binned and the count is represented by a rectangular bar. We choose an initial bin width of $1 and produce the uppermost graph of Fig. 1.1. The distribution appears to be unimodal; that is, it has one peak, the bar representing the tips greater than $1.50 and less than or equal $2.50. There are very few tips of $1.50 or less. The number of larger tips trails off rapidly, which suggests that this is not a very expensive restaurant.

The conclusions drawn from a histogram are often influenced by the choice of bin width, which is a parameter of the graph and not of the data. Figure 1.1 shows a histogram with a smaller bin width, $10c$. At the smaller bin width the shape is multimodal, and it is clear that there are large peaks at the full dollars and smaller peaks at the half dollar. This shows that the customers tended to round the tip to the nearest fifty cents or dollar.

This type of observation occurs frequently when studying histograms: A large bin width smooths out the graph and shows rough or global trends, whereas a smaller bin width highlights more local features. Since the bin width is an example of a graph parameter, experimenting with bin width is an example of exploring a set of related graphs. Exploring multiple related graphs can lead to insights that would not be apparent in any single graph.

So far we have not addressed the primary question: What relationships exist between tip and the other variables? Since the tip is usually calculated based on the bill, it is natural to look first at a graph of tip and bill. A common graph for looking at a pair of continuous variables is the scatterplot, as in Fig. 1.2. We see that the variables are highly correlated ($r = 0.68$), which confirms that tip is calculated from the bill. We have added a line representing a tip rate of 18%. Disappointingly for the waiter, there are many more points below the line than above it: There are many more "cheap tippers" than generous tippers. There are a couple of notable exceptions, especially one

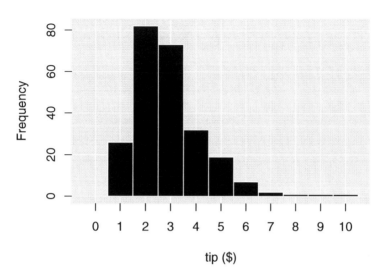

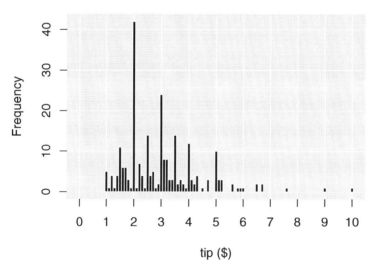

Fig. 1.1. Histograms of tip with differing bin width: $1, 10c. Bin width can be changed interactively in interactive systems, often by dragging a slider.

party who gave a $5.15 tip for a $7.25 bill, which works out to a tip rate of about 70%.

We said earlier that an essential aspect of data visualization is capturing relationships among many variables: three, four, or even more. This dataset, simple as it is, illustrates the point. Let us ask, for example, how a third variable such as sex affects the relationship between tip and bill. As sex is categorical with two levels (i.e., binary), it is natural to divide the data into female and male payers and to generate two scatterplots of tip vs. bill. Let us go even further by including a fourth variable, smoking, which is also binary. We now divide the data into four parts and generate the four scatterplots observed in Fig. 1.3. (The 18% tip guideline is included in each plot, and the correlation between the variables for each subset is in the top left of each plot.) Inspecting these plots reveals numerous features: (1) For smoking parties, there is a lot less association between the size of the tip and the size of the bill; (2) when a female non-smoker paid the bill, the tip was a very consistent percentage of the bill, with the exceptions of three dining parties; and (3) larger bills were mostly paid by men.

Taking stock

In the above example we gained a wealth of insight in a short time. Using nothing but graphical methods we investigated univariate, bivariate, and multivariate relationships. We found both global features and local detail. We saw that tips were rounded; then we saw the obvious correlation between the tip and the size of the bill, noting the scarcity of generous tippers; finally we discovered differences in the tipping behavior of male and female smokers and non-smokers.

Notice that we used very simple plots to explore some pretty complex relationships involving as many as four variables. We began to explore multivariate relationships for the first time when we produced the plots in Fig. 1.3. Each plot shows a subset obtained by partitioning the data according to two binary variables. The statistical term for partitioning based on variables is "conditioning." For example, the top left plot shows the dining parties that meet the condition that the bill payer was a male non-smoker: sex = male and smoking = False. In database terminology this plot would be called the result of "drill-down." The idea of conditioning is richer than drill-down because it involves a structured partitioning of *all* data as opposed to the extraction of a single partition.

Having generated the four plots, we arrange them in a two-by-two layout to reflect the two variables on which we conditioned. Although the axes in each plot are tip and bill, the axes of the overall figure are smoking (vertical) and sex (horizontal). The arrangement permits us to make several kinds of comparisons and to make observations about the partitions. For example, comparing the rows shows that smokers and non-smokers differ in the strength of the correlation between tip and bill, and comparing the plots in the top row shows that male and female non-smokers differ in that the larger bills tend

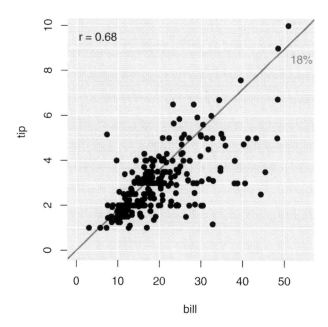

Fig. 1.2. Scatterplot of tip vs. bill. The line represents a tip of 18%. The greater number of points far below the line indicates that there are more "cheap tippers" than generous tippers.

to be paid by men. In this way a few simple plots allow us to reason about relationships among four variables.

In contrast, an old-fashioned approach without graphics would be to fit a regression model. Without subtle regression diagnostics (which rely on graphics!), this approach would miss many of the above insights: the rounding of tips, the preponderance of cheap tippers, and perhaps the multivariate relationships involving the bill payer's sex and the group's smoking habits.

1.4 Getting real: process and caveats

The preceding explanations may have given a somewhat misleading impression of the process of data analysis. In our account the data had no problems; for example, there were no missing values and no recording errors. Every step was logical and necessary. Every question we asked had a meaningful answer. Every plot that was produced was useful and informative. In actual data

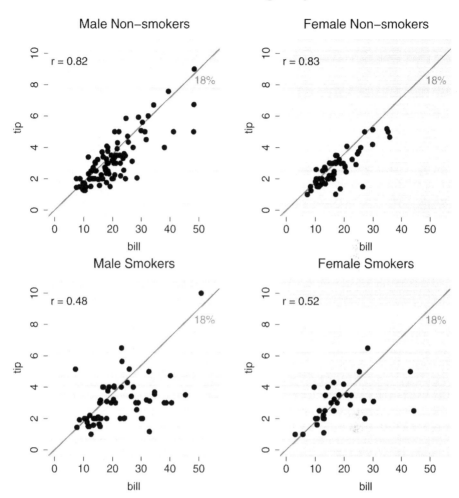

Fig. 1.3. Scatterplot of tip vs. bill conditioned by sex and smoker. There is almost no association between tip and bill in the smoking parties, and with the exception of three dining parties, when a female non-smoker paid the bill, the tip was extremely consistent.

analysis, nothing could be further from the truth. Real datasets are rarely perfect; most choices are guided by intuition, knowledge, and judgment; most steps lead to dead ends; most plots end up in the wastebasket. This may sound daunting, but even though data analysis is a highly improvisational activity, it can be given some structure nonetheless.

To understand data analysis, and how visualization fits in, it is useful to talk about it as a process consisting of several stages:

- The problem statement
- Data preparation
- Exploratory data analysis
- Quantitative analysis
- Presentation

The problem statement: Why do you want to analyze this data? Underlying every dataset is a question or problem statement. For the tipping data the question was provided to us from the data source: "What are the factors that affect tipping behavior?" This problem statement drives the process of any data analysis. Sometimes the problem is identified prior to a data collection. Perhaps it is realized after data becomes available because having the data available has made it possible to imagine new issues. It may be a task that the boss assigns, it may be an individual's curiosity, or it may be part of a larger scientific endeavor. Ideally, we begin an analysis with some sense of direction, as described by a pertinent question.

Data preparation: In the classroom, the teacher hands the class a single data matrix with each variable clearly defined. In the real world, it can take a great deal of work to construct a clean data matrix. For example, data values may be missing or misrecorded, data may be distributed across several sources, and the variable definitions and data values may be inconsistent across these sources. Analysts often have to invest considerable time in acquiring domain knowledge and in learning computing tools before they can even ask a meaningful question about the data. It is therefore not uncommon for this stage to consume most of the effort that goes into a project. And it is also not uncommon to loop back to this stage after completing the subsequent stages, to re-prepare and re-analyze the data.

In preparing the Tips data, we would create a new variable called tip rate, because when tips are discussed in restaurants, among waiters, dining parties, and tourist guides, it is in terms of a percentage of total bill. We may also create several new dummy variables for the day of the week, in anticipation of fitting a regression model. We did not talk about using visualization to verify that we had correctly understood and prepared the tipping data. For example, that unusually large tip could have been the result of a transcription error. Graphics identified the observation as unusual, and the analyst might use this information to search the origins of the data to check the validity of the numbers for this observation.

Exploratory data analysis (EDA): At this stage in the analysis, we make time to "play in the sand" to allow us to find the unexpected, and come to some understanding of our data. We like to think of this as a little like traveling. We may have a purpose in visiting a new city, perhaps to attend a conference, but we need to take care of our basic necessities, such as finding eating places and gas stations. Some of our movements will be pre-determined, or guided by the advice of others, but some of the time we wander around by ourselves.

1.4 Getting real: process and caveats   11

We may find a cafe we particularly like or a cheaper gas station. This is all about getting to know the neighborhood.

By analogy, at this stage in the analysis, we relax the focus on the problem statement and explore broadly different aspects of the data. Modern exploratory data analysis software is designed to make this process as fruitful as possible. It is a highly interactive, real-time, dynamic, and visual process, having evolved along with computers. It takes advantage of technology, in a way that Tukey envisioned and experimented with on specialist hardware 40 years ago: "Today, software and hardware together provide far more powerful factories than most statisticians realize, factories that many of today's most able young people find exciting and worth learning about on their own" (Tukey 1965). It is characterized by direct manipulation and dynamic graphics: plots that respond in real time to an analyst's queries and change dynamically to re-focus, link to information from other sources, and re-organize information. The analyst can work rapidly and thoroughly through the data, slipping out of dead-ends and chasing down new leads. The high level of interactivity is enabled by bare-bones graphics, which are generally not adequate for presentation purposes.

We gave you some flavor of this stage in the analysis of the waiter's tips. Although the primary question was about the factors affecting tipping behavior, we checked the distribution of individual variables, we looked for unusual records, we explored relationships among multiple variables, and we found some unexpected patterns: the rounding of tips, the prevalence of cheap tippers, and the heterogeneity in variance between groups.

Quantitative analysis (QA):

At this stage, we use statistical modeling and statistical interference to answer our primary questions. With statistical models, we summarize complex data, decomposing it into estimates of signal and noise. With statistical inference, we try to assess whether a signal is real. Data visualization plays an important role at this stage, although that is less well known than its key role in exploration. It is helpful both in better understanding a model and in assessing its validity in relation to the data.

For Tips, we have not yet answered the primary question of interest. Let's fit a regression model using tiprate as the response and the remaining variables (except tip and bill) as the explanatory variables. When we do this, only size has a significant regression coefficient, resulting in the model $\hat{tiprate} = 0.18 - 0.01 \times size$. The model says that, starting from a baseline tip rate of 18%, the amount drops by 1% for each additional diner in a party, and this is the model answer in Bryant & Smith (1995). Figure 1.4 shows this model and the underlying data. (The data is *jittered* horizontally to alleviate overplotting caused by the discreteness of size; that is, a small amount of noise is added to the value of size for each case.)

Are we satisfied with this model? We have some doubts about it, although we know that something like it is used in practice: Most restaurants today factor the tip into the bill automatically for larger dining parties. However, in this data it explains only 2% of the variation in tip rate. The points are spread widely around the regression line. There are very few data points for parties of size one, five, and six, which makes us question the validity of the model in these regions. The signal is very weak relative to the noise.

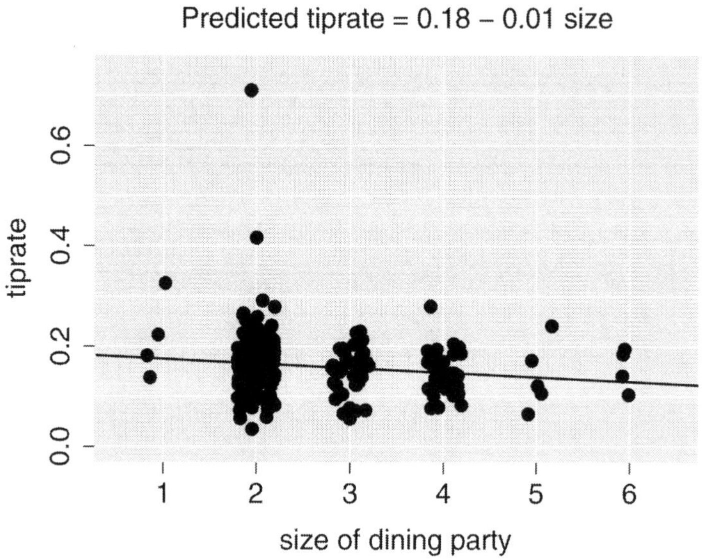

Predicted tiprate = 0.18 − 0.01 size

[scatterplot of tiprate vs. size of dining party, with regression line]

Fig. 1.4. Factors affecting tipping behavior. This scatterplot of **tiprate** vs. **size** shows the best model along with the data (jittered horizontally). There is a lot of variation around the regression line, showing very little signal relative to noise. In addition there are very few data points for parties of 1, 5, or 6 diners, so the model may not be valid at these extremes.

Most problems are more complex than the Tips data, and the models are often more sophisticated, so evaluating them is correspondingly more difficult. We evaluate a model using data produced by the model-fitting process, such as model estimates and diagnostics. Other data may be derived by simulating from the model or by calculating confidence regions. All this data can be explored and plotted for the pleasure of understanding the model.

Plotting the model in relation to the original data is also important. There is a temptation to ignore that messy raw data in favor of the simplification provided by a model, but a lot can be learned from what is left out of a model. For example, we would never consider teaching regression analysis without teaching residual plots. A model is a succinct explanation of the variation in the data, a simplification. With a model we can make short descriptive statements about the data, and pictures help us find out whether a model is *too* simple. And so we plot the model in the context of the data, as we just did in Fig. 1.4, and as we will do often in the chapters to follow.

The interplay of EDA and QA: Is it data snooping?

Because EDA is very graphical, it sometimes gives rise to a suspicion that patterns in the data are being detected and reported that are not really there. Sometimes this is called *data snooping*. Certainly it is important to validate our observations about the data. Just as we argue that models should be validated by all means available, we are just as happy to argue that observations made in plots should be validated using quantitative methods, permutation tests, or cross-validation, as appropriate, and incorporating subject matter expertise. A discussion of this topic emerged in the comments on Koschat & Swayne (1996), and Buja's remark (Buja 1996) is particularly apt:

> In our experience, false discovery is the lesser danger when compared to nondiscovery. Nondiscovery is the failure to identify meaningful structure, and it may result in false or incomplete modeling. In a healthy scientific enterprise, the fear of nondiscovery should be at least as great as the fear of false discovery.

We snooped into the Tips data, and from a few plots we learned an enormous amount of information about tipping: There is a scarcity of generous tippers, the variability in tips increases extraordinarily for smoking parties, and people tend to round their tips. These are very different types of tipping behaviors than what we learned from the regression model. The regression model was not compromised by what we learned from graphics, and indeed, we have a richer and more informative analysis. Making plots of the data is just smart.

On different sides of the pond: EDA and IDA

Consulting statisticians, particularly in the British tradition, have always looked at the data before formal modeling, and call it IDA (initial data analysis) (Chatfield 1995). For example, Crowder & Hand (1990) say: "The first thing to do with data is to look at them.... usually means tabulating and plotting the data in many different ways to 'see what's going on'. With the wide availability of computer packages and graphics nowadays there is no excuse for ducking the labour of this preliminary phase, and it may save some red faces later."

The interactive graphics methods described in this book emerged from a different research tradition, which started with Tukey's influential work on EDA, focusing on discovery and finding the unexpected in data. Like IDA, EDA has always depended heavily on graphics, even before the term *data visualization* was coined. Our favorite quote from John Tukey's rich legacy is that we need good pictures to "force the unexpected upon us."

EDA and IDA, although not entirely distinct, differ in emphasis. Fundamental to EDA is the desire to let the data inform us, to approach the data without pre-conceived hypotheses, so that we may discover unexpected features. Of course, some of the unexpected features may be errors in the data. IDA emphasizes finding these errors by checking the quality of data prior to formal modeling. It is much more closely tied to inference than EDA: Problems with the data that violate the assumptions required for valid inference need to be discovered and fixed early.

In the past, EDA and inference were sometimes seen as incompatible, but we argue that they are not mutually exclusive. In this book, we present some visual methods for assessing uncertainty and performing inference, that is, deciding whether what we see is "really there."

Presentation: Once an analysis has been completed, the results must be reported, either to clients, managers, or colleagues. The results probably take the form of a narrative and include quantitative summaries such as tables, forecasts, models, and graphics. Quite often, graphics form the bulk of the summaries.

The graphics included in a final report may be a small fraction of the graphics generated for exploration and diagnostics. Indeed, they may be different graphics altogether. They are undoubtedly carefully prepared for their audience. The graphics generated during the analysis are meant for the analyst only and thus need to be quickly generated, functional but not polished. This issue is a dilemma for authors who have much to say about exploratory graphics but need to convey it in printed form. The plots in this book, for example, lie somewhere between exploratory and presentation graphics.

As mentioned, these broadly defined stages do not form a rigid recipe. Some stages overlap, and occasionally some are skipped. The order is often shuffled and groups of steps reiterated. What may look like a chaotic activity is often improvisation on a theme loosely following the "recipe."

Because of its improvisational nature, EDA is not easy to teach. Says Tukey (1965) "Exploratory data analysis is NOT a bundle of techniques....Confirmatory analysis is easier to teach and compute...." In the classroom, the teacher explains a method to the class and demonstrates it on the single data matrix and then repeats this process with another method. Teaching a bundle of methods is indeed an efficient approach to covering substantial quantities of material, but this may be perceived by the student as a stream of disconnected methods, applied to unrelated data fragments, and they may not be able to apply what they have learned outside that fragmented context for quite a while. It takes time and experience for students to integrate this material and to develop their own intuition. Students need to navigate their own way through data, cleaning it, exploring it, choosing models; they need to make mistakes, recover from them, and synthesize the findings into a summary. Learning how to perform data analysis is a process that continues long after the student's formal training is complete.

1.5 Interactive investigation

Thus far, all observations on the tipping data have been made using static graphics — our purpose up to this point has been to communicate the importance of plots in the context of data analysis. Static plots were originally drawn by hand, and although they are now produced by computers, they are still designed to be printed on paper, often to be displayed or studied some time later. However, computers also allow us to produce plots to be viewed as they are created, and tweaked and manipulated in real time. This book is about such interactive and dynamic plots, and the chapters that follow have a lot to say about them. Here we will say a few words about the way interactive plots enhance the data analysis process we have just described.

The Tips data is simple, and most of the interesting features can be discovered using static plots. Still, interacting with the plots reveals more and enables the analyst to pursue follow-up questions. For example, we could address a new question, arising from the current analysis, such as "Is the rounding behavior of tips predominant in some demographic group?" To investigate we probe the histogram, highlight the bars corresponding to rounded tips, and observe the pattern of highlighting in the linked plots (Fig. 1.5). Multiple plots are visible simultaneously, and the highlighting action on one plot generates changes in the other plots. The two additional plots here are *mosaic plots*, which are used to examine the proportions in categorical variables. (Mosaic plots will be explained further in the next chapter; for now, it is enough to know that the area of each rectangle is proportional to the corresponding number of cases in the data.) For the highlighted subset of dining parties, the ones who rounded the tip to the nearest dollar or half-dollar, the proportion of bill paying males and females is roughly equal, but interestingly, the proportion of smoking parties is higher than non-smoking parties. This might suggest another behavioral difference between smokers and non-smokers: a larger tendency for smokers than non-smokers to round their tips. If we were to be skeptical about this effect we would dig deeper, making more graphical explorations and numerical models. By pursuing this with graphics, we would find that the proportion of smokers who round the tip is only higher than non-smokers for full dollar amounts, and not for half-dollar amounts.

The remaining chapters in this book continue in this vein, describing how interactive and dynamic plots are used in several kinds of data analysis.

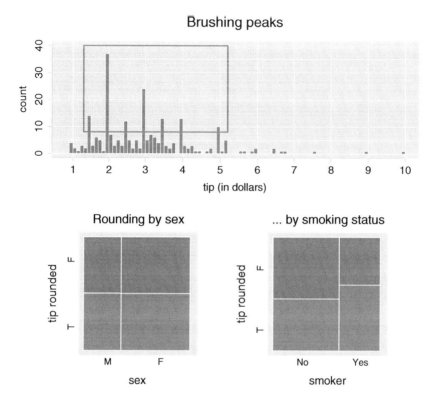

Fig. 1.5. Histogram of tip linked to 2D mosaic plots of sex and smoker. Bars of whole and half-dollar amounts are highlighted. The proportion of smoking parties who round their tips is higher than that of non-smoking parties, whereas men and women round about equally.

2
The Toolbox

The tools used to perform the analyses described in this book come largely from two "toolboxes." One is R, which we use extensively for data management and manipulation, modeling, and static plots. Since R is well documented elsewhere, both in books (Dalgaard 2002, Venables & Ripley 2002, Murrell 2005) and on the web (R-project.org), we will say very little about it here.

Instead, we emphasize the less familiar tools drawn from GGobi, our other major toolbox: a set of direct manipulations that we apply to a set of plot types. With these plot types and manipulations we can construct and interact with multiple views, linked so that an action in one can affect them all. The examples described throughout the book are based on these tools.

2.1 Introduction

It will be helpful to have a shorthand for describing the information that is used to generate a plot and what is shared between plots in response to user manipulations. We will introduce this notation using the Australian Crabs data, an excerpt of which is shown in Table 2.1.

The data table is composed of two categorical variables (species and sex) and five real-valued variables (frontal lobe, rear width, carapace length, carapace width, and body depth). The real-valued variables are the physical measurements taken on each crab, and the categorical variables are class or grouping labels for each crab. This distinction is important for visualization as well as for analysis, and it is made throughout the examples in the chapters to follow. Graphics for real-valued variables and categorical variables differ substantially, and they complement one another. For example, when both variable types are present, it is common to plot a pair of real-valued variables in a scatterplot and a single categorical variable in a barchart, and to color the points in the scatterplot according to their level in the categorical variable.

Table 2.1 Example data table, an excerpt from Australian Crabs

Crab	species	sex	frontal lobe	rear width	carapace length	carapace width	body depth
1	blue	male	8.1	6.7	16.1	19.0	7.0
2	blue	male	8.8	7.7	18.1	20.8	7.4
3	blue	male	9.2	7.8	19.0	22.4	7.7
4	blue	male	9.6	7.9	20.1	23.1	8.2
51	blue	female	7.2	6.5	14.7	17.1	6.1
52	blue	female	9.0	8.5	19.3	22.7	7.7
53	blue	female	9.1	8.1	18.5	21.6	7.7
101	orange	male	9.1	6.9	16.7	18.6	7.4
102	orange	male	10.2	8.2	20.2	22.2	9.0
151	orange	female	10.7	9.7	21.4	24.0	9.8
152	orange	female	11.4	9.2	21.7	24.1	9.7
153	orange	female	12.5	10.0	24.1	27.0	10.9

The variables in a data table can be described algebraically using a matrix having n observations and p variables denoted as:

$$\mathbf{X} = [\mathbf{X}_1\ \mathbf{X}_2\ \ldots\ \mathbf{X}_p] = \begin{bmatrix} X_{11} & X_{12} & \ldots & X_{1p} \\ X_{21} & X_{22} & \ldots & X_{2p} \\ \vdots & \vdots & & \vdots \\ X_{n1} & X_{n2} & \ldots & X_{np} \end{bmatrix}_{n \times p}$$

For some analyses, we consider real-valued variables separately. As a subset of the Crabs data, the values in the table form a 12×5 matrix, $\mathbf{X}_{12 \times 5} = [\mathbf{X}_1, \ldots, \mathbf{X}_5]$. We will use this shorthand notation when we discuss methods that are applied only to real-valued variables. It is particularly useful for describing tour methods, which are presented in Sect. 2.2.3.

For the Crabs data, we are interested in understanding the variation in the five physical (real-valued) variables, and whether that variation depends on the levels of the two categorical variables. In statistical language, we may say that we are interested in the *joint distribution* of the five physical measurements *conditional* on the two categorical variables. A plot of one column of numbers displays the *marginal distribution* of one variable. Similarly a plot of two columns of the data displays the marginal distribution of two variables. Ultimately we want to describe the distribution of observed values in the five-dimensional space of the physical measurements.

Building insight about structure in high-dimensional spaces starts simply. We start with univariate and bivariate plots, looking for low-dimensional structure, and work our way up to multivariate plots to seek relationships among several variables. Along the way, we use different approaches to explore real-valued and categorical variables.

With larger datasets it is convenient to store the data table in a database and to access it with database queries. For example, this SQL command to subset the Crabs data would return the frontal lobe values of the female Blue crabs:

```
SELECT frontal_lobe FROM Australian Crabs
WHERE species=blue AND sex=female
```

Database querying is a way to conceptually frame direct manipulation methods [see, for example, Ahlberg, Williamson & Shneiderman (1991) and Buja, McDonald, Michalak & Stuetzle (1991)]. The above SQL query can be constructed graphically using interactive brushing, and the response can be presented visually, as described in Sect. 2.3.1.

This chapter starts with plot types, and continues with a discussion of plot manipulation. The section on plot types is divided into subsections corresponding to the number of data dimensions displayed, and it ends with a brief discussion of the simultaneous use of multiple views. The section on direct manipulation describes brushing and labeling (often using linked views), interactive view scaling, and moving points. We will close the chapter by mentioning some interesting tools in other software packages.

We will focus on two datasets to demonstrate the tools: Australian Crabs and Tips (described in Chap. 1). Tips will be used to illustrate categorical variable plots, and Crabs will be used to illustrate real-valued variable plots and plots combining categorical and real-valued variables. A few other datasets will make brief appearances.

2.2 Plot types

2.2.1 Univariate plots

One-dimensional (1D) plots such as histograms, box plots, or dot plots are used to examine the marginal distributions of single variables. What shape do the values show — unimodal or multimodal, symmetric or skewed? Are there clumps or clusters of values? Are a few values extremely different from the rest? Observations like these about a data distribution can only be made by plotting the data.

Each univariate plot type uses one column of the data matrix, $\mathbf{X}_i, i = 1, \ldots, p$.

Real-valued variables

Two types of univariate plots for real-valued variables are regularly used in this book, each one a variant of the dot plot. In a dot plot, one row of data generates one point on the plot, explicitly preserving the identity of each observation. This is useful for linking information between plots using direct manipulation, which is discussed later in the chapter. We also use histograms, where bar height represents either the count or the relative frequency of values within a bin range; we usually use the count. These are examples of "area plots," in which observations are aggregated into groups and each group is represented by a single symbol (in this case, a rectangle). Area plots are especially useful when the number of points is large enough to slow down the computer's response time or to cause over-plotting problems.

The plots in Fig. 2.1 show three different univariate plots for the column X_3 (frontal lobe), for a subset of the data corresponding to the female crabs of the Blue species. The first plot is a histogram, and the next two are dot plots.

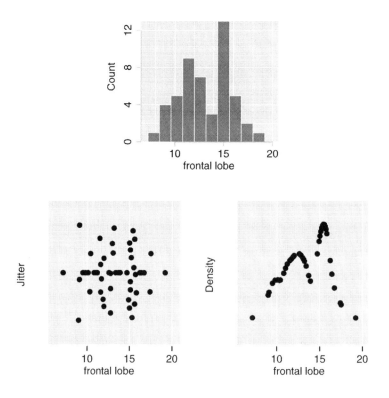

Fig. 2.1. Univariate displays: Histogram, textured dot plot, and average shifted histogram of frontal lobe of the female crabs of the Blue species. We see bimodality in the distribution of values, with many cases clustered near 15 and another cluster of values near 12.

In a dot plot, each case is represented by a small circle (i.e., a dot). Since there are usually several dots with the same or similar values, some method must be used to separate those dots. They may be *stacked*, which results in a plot that is much like a histogram, or they may be *jittered*, which adds a small amount of noise to each point. Jittering sometimes results in clumps of points that give the false impression of meaningful clustering, so the *textured dot plot* was developed (Tukey & Tukey 1990) to spread the data more evenly. When there are three or fewer cases with the same or similar data values, those points are placed at constrained locations; when there are more than three cases with the same value, the points are positioned using a combination of constraint and randomness.

The last plot in Fig. 2.1 is an *average shifted histogram* or *ASH* plot, devised by Scott (1992). In this method, several histograms are calculated using the same bin width but different origins, and the averaged bin counts at each data point are plotted. His algorithm has two key parameters: the number of bins, which controls the bin width, and the number of histograms to be computed. The effect is a smoothed histogram, and since it can shown as a dot plot, it is a histogram that can be linked case by case to other scatterplots.

Categorical variables

One-dimensional categorical data is commonly plotted in a bar chart, where the height of the bar is proportional to the number of cases in each category. Figure 2.2 displays a barchart of day in the Tips data: There were fewer diners on Friday than on the other three days for which we have data. An alternative to a bar chart is a spine plot, where the category count is represented by the width of the bar rather than by the height. When there is only one variable, the spine plot tells the same story as does the barchart; its special value emerges when plotting more than one categorical variable, as should become clear shortly.

2.2.2 Bivariate plots

Plots of two variables are used for examining the joint distribution of two variables. This pairwise relationship is often a marginal distribution of multivariate data.

Real-valued variables

We use two-dimensional (2D) scatterplots to see the relationship between two real-valued variables. This relationship may be linear, non-linear, or non-existent. We also look for deviations from dependence, such as outliers, clustering, or heterogeneous variation.

An example is shown in the two plots in Fig. 2.3. In the first plot, the two plotted variables are a subset of the five physical measurements of the female

22   2 The Toolbox

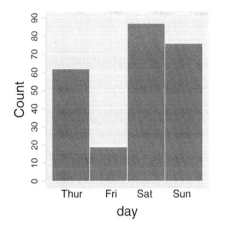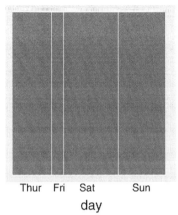

Fig. 2.2. Barchart and spine plot of day in the Tips data. There are relatively fewer records on Friday, as we can see in both plots. Friday's bar is the shortest one in the barchart **(left)** and the narrowest one in the spine plot.

Blue Australian Crabs. The plot shows a strong linear relationship between rear width and frontal lobe. In the second plot, we have shown a transformation of the data: the same two variables after they have been sphered using a principal component transformation, which has removed the linear relationship.

Scatterplots are often enhanced by overlaying density information using contours, color, or a grayscale. If one variable can be considered a response and the other an explanatory variable, we may add regression curves or smoothed lines.

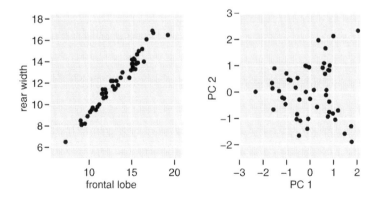

Fig. 2.3. Bivariate scatterplots for the Blue female Australian Crabs. The scatterplot of rear width against frontal lobe **(left)** shows a strong linear relationship. The plot of principal component 2 against principal component 1 shows that the principal component transformation removes the linear relationship.

Categorical variables

A spine plot can also be defined as a one-dimensional (1D) mosaic plot. When we add a second categorical variable, we can use the resulting 2D mosaic plot to explore the dependence between the two variables. The left-hand plot in Fig. 2.4 shows a mosaic plot of day and sex from the Tips data. The bars are split into gray and orange sections representing male and female bill payers.

The heights of the areas representing males increase from Thursday to Sunday, which shows the change in the relative proportion of male to female bill-payers. If this data were displayed as a stacked bar chart, as in the right-hand plot, it would be quite difficult to compare the relative proportions of each bar; that is, it would be hard to decide whether the proportion of female bill-payers varies with day of the week.

A mosaic plot can handle several categorical variables; it shows the frequencies in an n-way contingency table, each one corresponding to a rectangle with proportional area. It is constructed by repeatedly subdividing rectangular areas. In the example, the size of each rectangular area in the example is proportional to the number of cases having level i of the day and level j of sex. If we were to add a third categorical variable, such as smoker, we would divide each of the eight rectangles by a horizontal division.

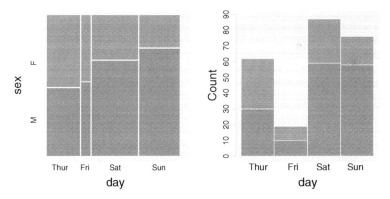

Fig. 2.4. A mosaic plot of day and sex for the Tips data, in which male and female bill payers are colored gray and orange, respectively. The proportion of male bill payers increases from Thursday to Sunday **(left)**, which is easier to read from this display than from a stacked barchart **(right)**.

2.2.3 Multivariate plots

Parallel coordinate plots for categorical or real-valued variables

Parallel coordinate plots (Inselberg 1985, Wegman 1990) are frequently used for examining multiple variables, looking for correlations, clustering in high dimensions, and other relationships. The plots are constructed by laying out the axes in parallel instead of in the more familiar orthogonal orientation of the Cartesian coordinate system. Cases are represented by a line trace connecting the case value on each variable axis. Mathematician d'Ocagne (1885) was the first to explain the geometry of a parallel coordinate plot, and how it is that a point on a graph of Cartesian coordinates transforms into a line in this other space.

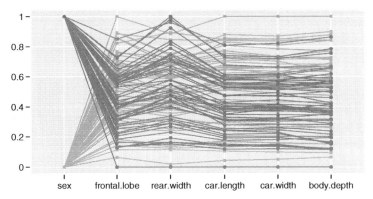

Fig. 2.5. Parallel coordinate plot of six of the seven variables of the Australian Crabs data, with males and females identified as green rectangles and purple circles, respectively. The relatively flat traces indicate strong correlation between variables.

Figure 2.5 shows a parallel coordinate plot of one categorical variable, sex, and the five physical measurement variables of the Blue species of the Australian Crabs. The female crabs are shown by purple circles, and the male crabs are shown by green rectangles. We first note that the trace for each crab is relatively flat, which indicates strong correlation between variables. (A small crab, for example, is small in all dimensions.) Second, the lines for males and females cross between frontal lobe, rear width, and carapace length, which suggests that the differences between males and females can be attributed to rear width.

The order in which the parallel axes are positioned influences the viewer's ability to detect structure. Ordering the layout by a numerical measure, such as correlation between variables, can be helpful. When the variables have a natural order, such as time in longitudinal or repeated measures, these plots are essentially the same as profile plots. Parallel coordinate plots are also the same as interaction plots, which are used in plotting experimental data containing several factors.

Scatterplot matrix, for categorical or real-valued variables

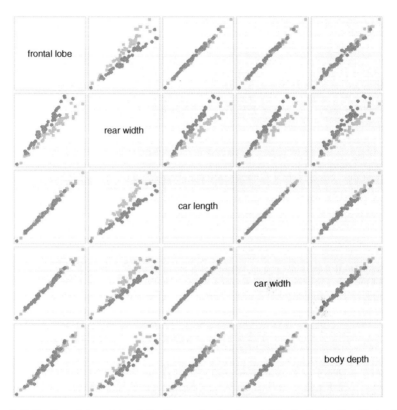

Fig. 2.6. Scatterplot matrix of the five real-valued variables in Australian Crabs. The scatterplot matrix is a display of related plots, in an arrangement that allows us to learn from seeing them in relation to one another. All five variables are strongly linearly related, but some structure can be observed.

The scatterplot matrix (draftsman's plot) contains pairwise scatterplots of the p variables laid out in a matrix format. It is a compact method for displaying a number of relationships at the same time, and it offers something more, because this sensible plot arrangement allows us to simultaneously make comparisons among all the plots. As with parallel coordinate plots, it is often useful to re-order variables to highlight comparisons.

Figure 2.6 displays a scatterplot matrix of the five physical measurement variables of the males and females for the Blue species of the Australian Crabs, and the diagonal displays the univariate ASH plot for each variable. (Once again, male and female crabs are represented by green rectangles and purple circles, respectively.) We can quickly see that all five variables are strongly linearly related. Note too that the two sexes differ in rear width more than they do on the other variables; females have a relatively larger value for rear width than males. The difference is more pronounced in larger crabs.

It is instructive to compare the scatterplot matrix to the correlation or covariance matrix for the same variables since they share the same square structure. For example, each plot in the scatterplot matrix just discussed corresponds to a single number in this correlation matrix:

	FL	RW	CL	CW	BD
FL	1.00	0.90	1.00	1.00	0.99
RW	0.90	1.00	0.90	0.90	0.90
CL	1.00	0.90	1.00	1.00	0.99
CW	1.00	0.90	1.00	1.00	0.99
BD	0.99	0.90	1.00	1.00	1.00

Just as one scatterplot contains a great deal more information than can be captured by any one statistic, a scatterplot matrix contains much more information than this table of statistics.

Tours for real-valued variables

A tour is a motion graphic designed to study the joint distribution of multivariate data (Asimov 1985), in search of relationships that may involve several variables. It is created by generating a sequence of low-dimensional projections of high-dimensional data; these projections are typically 1D or 2D. Tours are thus used to find interesting projections of the data that are not orthogonal, unlike those plotted in a scatterplot matrix and other displays of marginal distributions.

Here is a numerical calculation illustrating the process of calculating a data projection. For the five physical measurements in the crabs data, in Table 2.1, if the 1D projection vector is

$$\mathbf{A}_1 = \begin{bmatrix} 1 \\ 0 \\ 0 \\ 0 \\ 0 \end{bmatrix}, \quad \text{then the data projection is} \quad \mathbf{XA}_1 = \begin{bmatrix} 8.1 \\ 8.8 \\ 9.2 \\ 9.6 \\ 7.2 \\ \vdots \end{bmatrix},$$

which is equivalent to the first variable, frontal lobe. Alternatively, if the projection vector is

$$\mathbf{A}_2 = \begin{bmatrix} 0.707 \\ 0.707 \\ 0 \\ 0 \\ 0 \end{bmatrix}, \text{ then } \mathbf{XA}_2 = \begin{bmatrix} 0.707 \times 8.1 + 0.707 \times 6.7 = 10.5 \\ 0.707 \times 8.8 + 0.707 \times 7.7 = 11.7 \\ 0.707 \times 9.2 + 0.707 \times 7.8 = 12.0 \\ 0.707 \times 9.6 + 0.707 \times 7.9 = 12.4 \\ 0.707 \times 7.2 + 0.707 \times 6.5 = 9.7 \\ \vdots \end{bmatrix},$$

and the resulting vector is a linear combination of the first two variables, frontal lobe and rear width.

The corresponding histograms of the full set of crabs data are shown in Fig. 2.7, with the projection of the data onto $\mathbf{A}_1$ at left and the projection onto $\mathbf{A}_2$ at right. In the linear combination of two variables, $\mathbf{XA}_2$, a suggestion of bimodality is visible. The data is also more spread in this projection: The variance of the combination of variables is larger than the variance of frontal lobe alone.

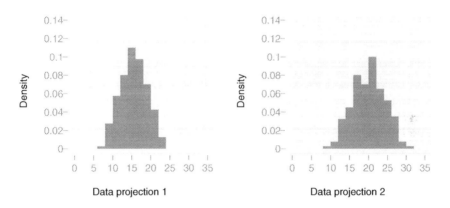

Fig. 2.7. Two 1D projections of the Crabs data. The plotted vectors are different linear combinations of frontal lobe and rear width, such that the first plot shows frontal lobe alone and the second shows an equal-weighted projection of both variables.

Next, we will look at a numerical example of a 2D projection. If the 2D projection matrix is

$$\mathbf{A}_3 = \begin{bmatrix} 1 & 0 \\ 0 & 1 \\ 0 & 0 \\ 0 & 0 \\ 0 & 0 \end{bmatrix}, \text{ then the data projection is } \mathbf{XA}_3 = \begin{bmatrix} 8.1 & 6.7 \\ 8.8 & 7.7 \\ 9.2 & 7.8 \\ 9.6 & 7.9 \\ 7.2 & 6.5 \\ \vdots \end{bmatrix},$$

which is equivalent to the first two variables, frontal lobe and rear width. Alternatively, if the projection matrix is

$$\mathbf{A}_4 = \begin{bmatrix} 0 & 0 \\ 0 & 0.950 \\ 0 & -0.312 \\ -0.312 & 0 \\ 0.950 & 0 \end{bmatrix} \quad \text{then} \quad \mathbf{XA}_4 =$$

$$\begin{bmatrix} -0.312 \times 19.0 + 0.950 \times 7.0 = 0.72 & 0.950 \times 6.7 - 0.312 \times 16.1 = 1.34 \\ -0.312 \times 20.8 + 0.950 \times 7.4 = 0.54 & 0.950 \times 7.7 - 0.312 \times 18.1 = 1.67 \\ -0.312 \times 22.4 + 0.950 \times 7.7 = 0.33 & 0.950 \times 7.8 - 0.312 \times 19.0 = 1.48 \\ -0.312 \times 23.1 + 0.950 \times 8.2 = 0.58 & 0.950 \times 7.9 - 0.312 \times 20.1 = 1.23 \\ -0.312 \times 17.1 + 0.950 \times 6.1 = 0.46 & 0.950 \times 6.5 - 0.312 \times 14.7 = 1.59 \\ \vdots & \end{bmatrix}.$$

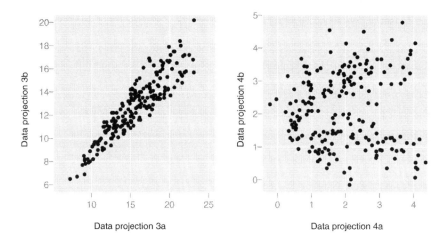

Fig. 2.8. Two 2D projections of the Crabs data. The first projection shows a strong linear relationship, and the second projection shows clustering within a narrower range of values.

The resulting matrix has two columns, the first of which is a linear combination of carapace width and body depth, and the second of which is a linear combination of rear width and carapace length.

The corresponding bivariate scatterplots of the full set of crabs data are shown in Fig. 2.8. The projection of the data into $\mathbf{A}_3$ is shown at left, and the projection into $\mathbf{A}_4$ is on the right. The left-hand projection shows shows a strong linear relationship, with larger variation among the high values. In the

2.2 Plot types 29

second projection the points loosely clump into the four sectors of the plot, and by inspecting the axes we can see that the variation in values is much smaller in this projection.

The values in a projection matrix, **A**, can be any values on $[-1, 1]$ with the constraints that the squared values for each column sum to 1 (normalization) and the inner product of two columns sums to 0 (orthogonality).

A tour is constructed by creating a sequence of projection matrices, $\mathbf{A}_t$. The sequence should be dense in the space, so that all possible low-dimensional projections are equally likely to be chosen in the shortest amount of time. Between each pair of projections in the sequence an interpolation is computed to produce a sequence of intermediate projection matrices — the result is a sense of continuous motion when the data projections are displayed. Watching the sequence is like watching a movie, except that we usually watch in real time rather than watching a stored animation. The algorithm is described in detail in Buja, Cook, Asimov & Hurley (2005), and Cook, Lee, Buja & Wickham (2006).

An alternative to motion graphics is available if the projection dimension is 1. In that case, the projections can be drawn as tour curves, similar to Andrews curves (Andrews 1972).

How do we choose the projections to plot? There are three methods for choosing projections, and they might all be used at different points in an analysis:

- The grand tour, in which the sequence of projections is chosen randomly. It is designed to cover the space efficiently, and it is used to get acquainted with the data.
- The projection pursuit guided tour, in which the sequence is guided by an algorithm in search of "interesting" projections. It is used to aid in the search for specific patterns such as clusters or outliers, depending on the projection pursuit function chosen by the user.
- Manual manipulation, in which the user chooses the sequence. It is often used to explore the neighborhood around a projection that has been reached using a random or guided tour, perhaps in search of an even better projection.

Grand tour: In the grand tour, the default method in GGobi, a random sequence of projections is displayed. It may be considered an interpolated random walk over the space of all projections. This method used here was discussed originally in Asimov (1985) and Buja & Asimov (1986). Related work on tours can be found in Wegman (1991), Tierney (1991), and Wegman, Poston & Solka (1998).

Random projections can be generated efficiently by sampling from a multivariate normal distribution. For a 1D projection vector having p elements, we start by sampling p values from a standard univariate normal distribution;

the resulting vector is a sample from a standard multivariate normal. We standardize this vector to have length 1, and the result is a random value on a $(p-1)$D sphere: This is our projection vector. For a 2D random projection of p variables, follow this procedure twice, and orthonormalize the second vector on the first one.

Guided tour: In a projection pursuit guided tour (Cook, Buja, Cabrera & Hurley 1995), the next target basis is selected by optimizing a function that specifies some pattern or feature of interest; that is, the sequence of projections, in the high-dimensional data space, is moved toward low-dimensional projections that have more of this feature.

Using projection pursuit (PP) in a static context, such as in R or S-PLUS, yields a number of static plots of projections that are deemed interesting. Combining projection pursuit with motion offers the interesting views in the context of surrounding views, which allows better structure detection and interpretation.

It works by optimizing a criterion function, called the projection pursuit index, over all possible dD projections of pD data,

$$\max f(\mathbf{XA}) \quad \forall \mathbf{A},$$

which are subject to the orthonormality constraints on $\mathbf{A}$.

The projection pursuit indexes in our toolbox include holes, central mass, LDA, and PCA (1D only). For an $n \times d$ matrix of projected data, $\mathbf{y} = \mathbf{XA}$, these indexes are defined as follows:

Holes

$$I_{\text{holes}}(\mathbf{A}) = \frac{1 - \frac{1}{n}\sum_{i=1}^{n} \exp(-\frac{1}{2}\mathbf{y}_i \mathbf{y}_i')}{1 - \exp(-\frac{p}{2})}$$

Central mass

$$I_{\text{CM}}(\mathbf{A}) = \frac{\frac{1}{n}\sum_{i=1}^{n} \exp(-\frac{1}{2}\mathbf{y}_i \mathbf{y}_i') - \exp(-\frac{p}{2})}{1 - \exp(-\frac{p}{2})}$$

PCA Defined for 1D projections, $d = 1$.

$$I_{\text{PCA}} = \frac{1}{n}\sum_{i=1}^{n} \mathbf{y}_i^2$$

LDA

$$I_{\text{LDA}}(\mathbf{A}) = 1 - \frac{|\mathbf{A}'\mathbf{W}\mathbf{A}|}{|\mathbf{A}'(\mathbf{W}+\mathbf{B})\mathbf{A}|}$$

where $\mathbf{B} = \sum_{i=1}^{g} n_i(\bar{\mathbf{y}}_{i.} - \bar{\mathbf{y}}_{..})(\bar{\mathbf{y}}_{i.} - \bar{\mathbf{y}}_{..})'$, $\mathbf{W} = \sum_{i=1}^{g}\sum_{j=1}^{n_i}(\mathbf{y}_{ij} - \bar{\mathbf{y}}_{i.})(\mathbf{y}_{ij} - \bar{\mathbf{y}}_{i.})'$ are the between- and within-group sum of squares matrices in a linear discriminant analysis, with g =number of groups, and $n_i, i = 1,....g$ is the number of cases in each group.

These formulas are written assuming that $\mathbf{X}$ has a mean vector equal to zero and a variance–covariance matrix equal to the identity matrix; that is, it has been sphered (transformed into principal component scores) or equivalently transformed into principal component coordinates. An implementation can incorporate these transformations on the fly, so sphering the data ahead of PP is not necessary. However, in our experience all of the projection pursuit indexes find interesting projections more easily when the data is sphered first.

The holes and central mass indexes (Cook, Buja & Cabrera 1993) derive from the normal density function. The first is sensitive to projections with few points in the center of the projection, and the second to those with a lot of points in the center. The LDA index (Lee, Cook, Klinke & Lumley 2005) derives from the statistics for MANOVA, and it is maximized when the centers of pre-defined groups in the data (indicated by symbol) are farthest apart. The PCA index derives from principal component analysis, and it finds projections where the data is most spread. Figures 2.9 and 2.10 show a few projections found using projection pursuit guided tours on the Crabs data.

Optimizing the projection pursuit index is done by a derivative-free random search method. A new projection is generated randomly. If the projection pursuit index value is larger than the current value, the tour moves to this projection. A smaller neighborhood is searched in the next iteration. The possible neighborhood of new projections continues to shrink, until no new projection can be found where the projection pursuit index value is higher than that of the current projection. At that point, the tour stops at that local maximum of the projection pursuit index. To continue touring, the user needs to break out of the optimization, either reverting to a grand tour or choosing a new random start.

Manual tour: In order to explore the immediate neighborhood around a particular projection, we pause the automatic motion of the tour and go to work using the mouse. One variable is designated as the "manipulation variable," and the projection coefficient for this variable is controlled by mouse movements; the transformation of the mouse movements is subject to the orthonormality constraints of the projection matrix. In GGobi, manual control is available for 1D and 2D tours.

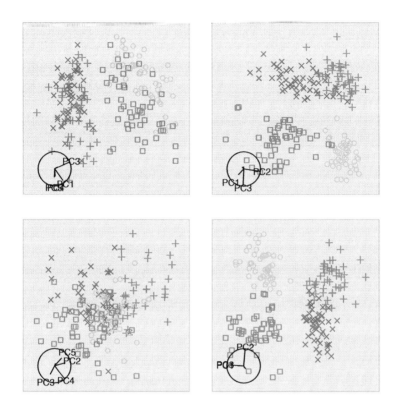

Fig. 2.9. Two-dimensional projections found by projection pursuit on the Crabs data. Two projections from the holes index (**top row**) show separation between the four classes. A projection from the central mass index (**bottom left**) shows the concentration of points in the center of the plot. A projection from the LDA index (**bottom right**) reveals the four classes.

In 1D tours the manipulation variable is rotated into or out of the current 1D projection.

In 2D tours, GGobi offers unconstrained oblique manipulation or constrained manipulation; if the manipulation is constrained, the permitted motion may be horizontal, vertical, or angular. With only three variables, the manual control works like a trackball: There are three rigid, orthonormal axes, and the projection is rotated by pulling the lever belonging to "manip" variable.

With four or more variables, a 3D manipulation space is created from the current 2D projection augmented by a third axis controlled by the "manip" variable. This manipulation space results from orthonormalizing the current projection with the third axis. When the "manip" variable is dragged around using the mouse, its coefficients follow the manual motion, and the coefficients

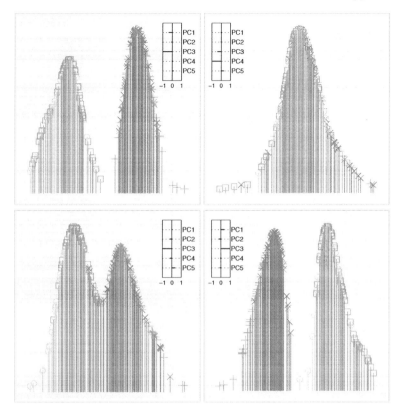

Fig. 2.10. One-dimensional projections found by projection pursuit on the Crabs data. A projection from the holes index has found separation by species (**top left**). A projection from the central mass index shows a density with short tails (**top right**); not especially useful for this data). The PCA index (**bottom left**) result shows bimodality but no separation. A projection from the LDA index (**bottom right**) reveals the species separation.

of the other variables are adjusted accordingly, such that their contributions to the 3D manipulation space are maintained.

When using the manual tour, we experiment with the choice of manipulation variable. We pause the tour to explore some interesting structure, and we select a manipulation variable. We may select a variable at random or make our selection based on our understanding of the data. We vary the variable's contribution to the projection and watch what happens, assessing the sensitivity of the structure to that variation. If we are lucky, we may be able to sharpen or refine a structure first exposed by the grand or guided tour.

Relationships between tours and numerical algorithms:

The tour algorithm is a cousin to several numerical algorithms often used in statistical data analysis — algorithms that generate interesting linear combinations of data. These linear combinations might be fed into a tour, or re-created directly using manual manipulation, after which their neighborhoods can be explored to see whether some nearby projection is even more informative.

In principal component analysis, for example, the principal component is defined to be $\mathbf{Y} = (\mathbf{X} - \bar{\mathbf{X}})\mathbf{A}$, where $\mathbf{A}$ is the matrix of eigenvectors from the eigen-decomposition of the variance–covariance matrix of the data, $\mathbf{S} = \mathbf{A}\Lambda\mathbf{A}'$. Thus a principal component is one linear projection of the data and could be one of the projections shown by a tour. A biplot (Gabriel 1971, Gower & Hand 1996) is a scatterplot that is similar to some 2D tour views: Plot the first two principal components, and then add the coordinate axes, which are analogous to the "axis tree" that is added at the lower left corner of 2D tour plots in GGobi. These axes are used to interpret any structure visible in the plot in relation to the original variables. Figure 2.11 shows biplots of the Australian crabs above a tour plot showing a similar projection, which is constructed using manual controls.

In linear discriminant analysis, the Fisher linear discriminant (the linear combination of the variables that gives the best separation of the class means with respect the class covariance) is one projection of the data that may be shown in a tour. In support vector machines (Vapnik 1999), projecting the data into the normal to the separating hyperplane yields another. Canonical correlation analysis and multivariate regression also generate interesting projections; these may be explored in a 2×1D tour.

2.2.4 Plot arrangement

Viewing multiple plots simultaneously is a good way to make comparisons among several variables. Examples include scatterplot matrices, parallel coordinate plots, and trellis plots (Becker, Cleveland & Shyu 1996). In each of these, several plots are arranged in a structured manner.

A scatterplot matrix lays out scatterplots of all pairs of the variables in a format matching a correlation or covariance matrix, which allows us to examine 2D marginal distributions in relation to one another. A parallel coordinate plot lays out all univariate dot plots in a sequential manner, which shows all 1D marginal distributions. The connecting lines give us more information about the relationships between variables, the joint distribution.

The arrangement of a trellis plot is different. The data is usually partitioned into different subsets of the cases, with each subset plotted separately. The partitioning is commonly based on the values of a categorical variable or on a binned real-valued variable. In statistical terms, we use trellis plots to examine the conditional distributions of the data. Brushing, which is described in the next section (2.3.1), can provide similar information.

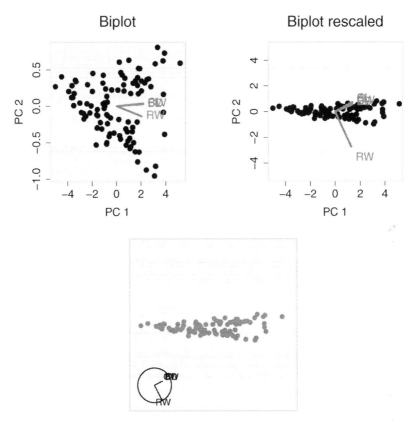

Fig. 2.11. The relationship between a 2D tour and the biplot: a biplot (**top left**) of the five physical measurement variables of the Australian Crabs; the same plot scaled to equal length axes (**top right**); the same projection (**bottom**) produced using the manually controlled tour.

2.3 Plot manipulation and enhancement

2.3.1 Brushing

Brushing works by using the mouse to control a "paintbrush," directly changing the color or glyph of elements of a plot. The paintbrush is sometimes a pointer and sometimes works by drawing an outline of sorts around points; the outline is sometimes irregularly shaped, like a lasso. In GGobi, the brush is rectangular when it is used to brush points in scatterplots or rectangles in area plots, and it forms a crosshair when it is used to brush lines.

Brushing is most commonly used when multiple plots are visible and some linking mechanism exists between the plots. There are several different conceptual models for brushing and a number of common linking mechanisms.

Brushing as database query

We can think about linked brushing as a query of the data. Two or more plots showing different views of the same dataset are near one another on the computer screen. A user poses a query graphically by brushing a group of plot elements in one of the plots. The response is presented graphically as well, as the corresponding plot elements in linked plots are modified simultaneously. A graphical user interface that supports this behavior is an implementation of the notion of "multiple linked views," which is discussed in Ahlberg et al. (1991) and Buja et al. (1991).

To illustrate, we set up two pairwise scatterplots of data on Australian Crabs, frontal lobe vs. rear width and sex vs. species. (In the latter plot, both variables are jittered.) These plots, shown in Fig. 2.12, are linked, case to case, point to point. We brush the points at the upper left in the scatterplot of sex vs. species, female crabs of the Blue species, and we see where those points appear in the plot of frontal lobe vs rear width. If we were instead working with the same data in a relational table in a database, we might pose the same question using SQL. We would issue the following simple SQL query and examine the table of values of frontal lobe and rear width that it returns:

```
SELECT frontal_lobe, rear_width FROM Australian_Crabs
WHERE sex = female AND species = Blue
```

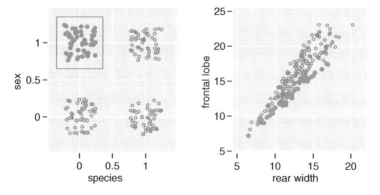

Fig. 2.12. A dynamic query posed using linked brushing. The points corresponding to sex = female and species = Blue are highlighted.

2.3 Plot manipulation and enhancement

Brushing is a remarkably efficient and effective way to both pose the query and view the result, allowing easy comparison of the values in the resulting table to the remaining values. Note that the plots containing query and response should be placed close together, so that the action occurs in the same visual neighborhood.

You may have already realized that the same task can be performed using a single window by first marking the queried objects and then selecting new variables to plot. Indeed, that approach has its uses, but it is much slower and the user may lose context as the view changes. Having several views available simultaneously allows the information to be rapidly absorbed.

A limitation of replacing SQL with brushing is that it may be difficult to form clauses involving many variables or otherwise complicated selections. Because most brushing interfaces use an accumulating mode of painting, it is easy to form disjunctions (unions) of any kind, but it may be unintuitive to form conjunctions (intersections) involving many variables. For such queries, a command line interface may be more efficient than extensive graphical control.

Brushing as conditioning

Since brushing a scatterplot defines a clause of the form $(X, Y) \in B$, the selection can be interpreted as conditioning on the variables X and Y. As we vary the size and position of the area B, we examine the dependence of variables plotted in the linked windows on the "independent" variables X and Y. Thus linked brushing allows the viewer to examine the conditional distributions in a multivariate dataset, and we can say that Fig. 2.12 shows the conditional distribution of frontal lobe and rear width on $X =$ species and $Y =$ sex.

This idea underlies scatterplot brushing described in Becker & Cleveland (1988) and forms the basis for trellis plots (Becker et al. 1996).

Brushing as geometric sectioning

Brushing projections with thin point-like or line-like brushes can be seen as forming geometric sections with hyperplanes in data space (Furnas & Buja 1994). From such sections the true dimensionality of the data can be inferred under certain conditions.

Linking mechanisms

Figure 2.12 shows the simplest kind of linking, one-to-one, where both plots show different projections of the same data, and a point in one plot corresponds to exactly one point in the other. When using area plots, brushing any part of an area has the same effect as brushing it all and is equivalent to selecting all cases in the corresponding category. Even when some plot elements

38   2 The Toolbox

represent more than one case, the underlying linking rule still links one case in one plot to the same case in other plots.

It is possible, though, to define a number of different linking rules. Your dataset may include more than one table of different dimensions, which are nevertheless related to one another through one or more of their variables. For example, suppose your dataset on crabs included a second table of measurements taken on Alaskan crabs. Instead of 200 crabs, as the first table describes, suppose the second table had data on only 60 crabs, with a few — but not all — of the same variables. You still might like to be able to link those datasets and brush on sex so that you could compare the two kinds of crabs. To establish that link, some rule is needed that allows m points in one plot to be linked to n points in another.

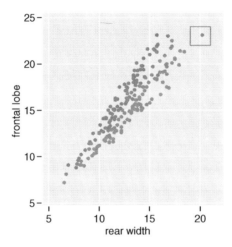

Fig. 2.13. Categorical brushing. Only a single point in the scatterplot of frontal lobe vs. rear width is enclosed by the brush, but because the linking rule specifies linking by sex, all points with sex = female have been brushed.

Figure 2.13 shows a variation we call linking by variable, or categorical brushing, in which a categorical variable is used to define the linking rule. It is used in plots of real-valued variables when the data also includes one or more categorical variables. Only a single point in the scatterplot is enclosed by the brush, but because the linking rule specifies linking by sex, all points with the same level of sex as the brushed point have been brushed simultaneously. In this simple example, $m = n$ and this takes place within a single plot. When we link two matrices of different dimension that share a categorical variable, we have true m to n linking.

Note that this distinction, linking by case or by variable, is moot for area plots. When the brush intersects an area, the entire area is brushed, so all cases with the corresponding levels are brushed — in that plot and in any linked plots. Linking by variable enables us to achieve the same effect within point plots.

Figure 2.14 illustrates another example of m-to-n linking. The data (Wages) contains 6,402 longitudinal measurements for 888 subjects. We treat the subject identifier as a categorical variable, and we specify it as our linking variable. When we brush any one of a subject's points, all other points corresponding to this subject are simultaneously highlighted — and in this instance, all connecting line segments between those points as well. In this case, we link m points in the dataset with n edges.

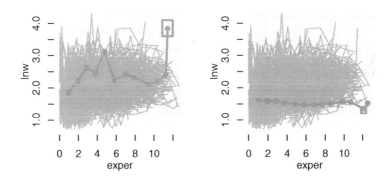

Fig. 2.14. Linking by subject identifier in longitudinal data, Wages, for two different subjects. The linking rule specifies linking by subject ID, so brushing one point causes all m points and n edges for that subject to be highlighted.

As we have just illustrated, we can link different kinds of objects. We can brush a point in a scatterplot that is linked to an edge in a graph. Figure 2.15 illustrates this using the Aradopsis Gene Expression data. The left plot shows the p-values vs. the mean square values from factor 1 in an analysis of aariance (ANOVA) model; it contains 8,297 points. The highlighted points are cases that have small p-values but large mean square values; that is, there is a lot of variation, but most of it is due to the treatment. The right plot contains 16,594 points that are paired and connected by 8,297 line segments. One line segment in this plot corresponds to a point in the left-hand scatterplot.

Persistent vs. transient — painting vs. brushing

Brushing scatterplots can be a transient operation, in which points in the active plot only retain their new characteristics while they are enclosed or intersected by the brush, or it can be a persistent operation, so that points retain their new appearance after the brush has been moved away. Transient

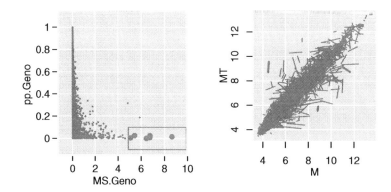

Fig. 2.15. Linking between a point in one plot and a line in another, using Arabidopsis Gene Expression. The line segments in the right-hand plot correspond exactly to the points in the left-hand plot.

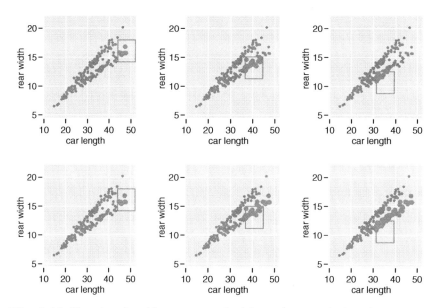

Fig. 2.16. Transient brushing contrasted with persistent painting. A sequence of transient operations (**top row**), in which the points return to their previous color when the brush moves off. In persistent painting (**bottom row**), points retain the new color.

brushing is usually chosen for linked brushing, as we have just described. Persistent brushing is useful when we want to group the points into clusters and then proceed to use other operations, such as the tour, to compare the groups. It is becoming common terminology to call the persistent operation "painting," and to reserve "brushing" to refer to the transient operation.

Painting can be done automatically, by an algorithm, rather than relying on mouse movements. The automatic painting tool in GGobi will map a color scale onto a variable, painting all points with the click of a button. (The GGobi manual includes a more detailed explanation.) Using the rggobi package, the setColors and setGlyphs commands can be used to paint points.

2.3.2 Identification

Identification, which could also be called labeling or label brushing, is another plot manipulation that can be linked. Bringing the cursor near a point or edge in a scatterplot, or a bar in a barchart, causes a label to appear that identifies the plot element. An identification parameter can be set that changes the nature of the label. Figure 2.17 illustrates different attributes shown using identification in GGobi: row label, variable value, and record identifier. The point highlighted is a female crab of the Orange species, with a value of 23.1 for frontal lobe, and it is the 200th row in the data matrix. Generally, identifying points is best done as a transient operation, but labels can be made to stick persistently. You will not be tempted to label many points persistently at the same time, because the plot will be too busy.

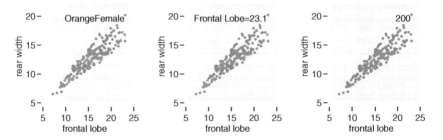

Fig. 2.17. Identifying points in a plot using three different labeling styles. The same point is labeled with its row label, variable value, or record identifier.

2.3.3 Scaling

Scaling maps the data onto the window, and changes in that mapping function help us learn different things from the same plot. Scaling is commonly used

42   2 The Toolbox

to zoom in on crowded regions of a scatterplot, and it can also be used to change the aspect ratio of a plot, to reveal different features of the data.

In the simple time series example shown in Fig. 2.18, we look at average monthly river flow plotted against time. The left plot shows the usual default scaling: a one-to-one aspect ratio. We have chosen this aspect ratio to look for global trends. An eye for detail might spot a weak trend, up and then down, and some large vertical gaps between points at the middle top.

The vertical gap can be seen even more explicitly if we zoom in (as you can try on your own), and there is a likely physical explanation: a sudden increase in river flow due to spring melt, followed by an equally sudden decline.

In the right plot, the aspect ratio has been changed by shrinking the vertical scale, called *banking to 45°* by Cleveland (1993). This reveals the seasonal nature of the river flow more clearly, up in the spring and down in the summer and fall.

As you can see, no single aspect ratio reveals all these features in the data equally well. This observation is consistent with the philosophy of exploratory data analysis, as distinct from a completely algorithmic or model-based approach, which may seek to find the "optimal" aspect ratio.

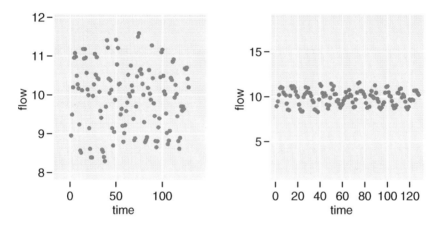

Fig. 2.18. Scaling a plot reveals different aspects of the River Flow data. At the default aspect ratio, a weak global trend is visible, up then down, as well as vertical gaps at the top. Flattening the plot highlights its seasonality.

2.3.4 Subset selection

Data analysts often partition or subset a data matrix and work with a portion of the cases. This strategy is helpful for several reasons. Sometimes the number of records is simply too large, and it slows down the computational software and causes a scatterplot to appear as an undifferentiated blur. If the data is fairly homogeneous, having no obvious clusters, a random sample of cases can be selected for testing and exploration. We can sample the cases in R, before handing the data to GGobi, or in GGobi itself, using the Subset tool.

If the data is inhomogeneous, even when the number of cases is not large, we may partition the cases in some structured way. One common partitioning strategy is to segment the data according to the levels of a categorical variable, creating groups in the data that can be analyzed and modeled one at a time. In that case, we may partition the cases in R before handing it to GGobi, or we may use brushing. We can paint the cases according to the levels of the partitioning variable, and then use the Color and glyph groups tool to show only the group we are currently studying.

If no categorical variable is available, we may choose one or more real-valued partitioning variables, dividing up the data according to their values.

2.3.5 Line segments

Line segments have already been mentioned as an important element of any display of longitudinal data. We also use them to illustrate and evaluate models, enhancing scatterplots with lines or grids, as you will see often in the chapters to follow. (See Figs. 5.4 and 5.10.)

In network analysis, the data includes a graph: a set of nodes connected by edges. In a social network, for example, a node often represents a person, and an edge represents the connection between two people. This connection may have associated categorical variables characterizing the connection (kinship, sexual contact, email exchanges, etc.) and real-valued variables capturing the frequency or intensity of the connection. Using GGobi, we can lay out graphs in a space of any dimension and use linked brushing to explore the values corresponding to both nodes and edges.

2.3.6 Interactive drawing

We sometimes enhance plots of data by adding points and lines to create illustrations of the model. (See Fig. 5.3.) Such illustrations can be prepared in R, or by using a text editor to edit the XML® data file, or they can be added interactively, using the mouse. The same tools allow us to edit graphs by hand.

2.3.7 Dragging points

Moving points is a very useful supplement to graph layout: No layout is perfect, and occasionally a point or group of points must be moved in order to clarify a portion of the graph.

You move a data point at your own peril, of course!

2.4 Tools available elsewhere

Many valuable tools are not currently in the GGobi toolbox but available in other software. Here we describe a few of these, and the software where the reader can find them.

- *Mosaic plots:* When two or more categorical variables are to be plotted, a good choice is a mosaic plot (Hartigan & Kleiner 1981) or double-decker plot (Hofmann 2001). These displays recursively partition a rectangle, with each rectangle representing the number of cases in the corresponding category; a good description can be found in Hofmann (2003). Direct manipulation includes the ability to manage the nesting order of variables in multivariate mosaic plots, and to interactively change the order in which the levels of categorical variables are displayed. They are also efficient plots for large data, since only a few rectangles may represent many thousands of points. Graphics for large data are extensively discussed in Unwin et al. (2006) and rich implementations of mosaic plots are included in MANET (Unwin, Hawkins, Hofmann & Siegl 1996) and Mondrian (Theus 2002).
- *Selection sequences:* If only two colors are used in a plot, one for background and one for highlighting, then brushing can be extended using *selection sequences* (Hofmann & Theus 1998). One subset selected by a brush can be combined with another brushed subset using Boolean logic. This tool is available in MANET (Unwin et al. 1996) and Mondrian (Theus 2002).
- *Alpha blending:* When many cases are plotted in a scatterplot or parallel coordinate plot, over-plotting is often a problem. One way to to see beneath the top point or line is to use *alpha blending*, where ink is laid down in transparent quantities. The more points drawn at the same location, the darker the spot becomes. Alpha blending is discussed in Carr et al. (1996), and a variant is used in XmdvTool (Fua, Ward & Rundensteiner 1999). Alpha levels can be adjusted interactively in Mondrian and in the R package `iPlots` (Urbanek & Theus 2003).
- *Brushing lines in a parallel coordinate plot:* Brushing in parallel coordinate plots is enhanced if it is possible to select lines as well as points. This is discussed in Carr et al. (1996) and explored in CASSATT (Unwin, Volinsky & Winkler 2003), and it is supported in MANET (Unwin et al. 1996) and Mondrian (Theus 2002). XmdvTool also has good support for parallel coordinate plots.
- *Maps:* A lot of data arises in a spatial context, and it is invaluable to show the map along with the data. See for example Anselin & Bao (1997), and Cook, Majure, Symanzik & Cressie (1996). GeoVista Studio (Takatsuka & Gahegan 2002) is a current development environment for creating software for exploring geospatial data, and ESTAT is a tool it has been used to build. In addition to scatterplots and highly interactive parallel coordinate plots, it includes linked displays of maps.

2.5 Recap

In this short introduction to our toolbox, we have not described all the tools in GGobi. Others are documented in the manual on the web site, which will change as the software evolves. The collection of methods in the toolbox is a very powerful addition for exploratory data analysis. We encourage the reader to explore the methods in more depth than is covered in this book, develop new approaches using rggobi, and explore the tools available in other software.

Exercises

1. For the Flea Beetles:
   a) Generate a parallel coordinate plot of the six variables, with the color of the lines coded according to species. Describe the differences between species.
   b) Rearrange the order of the variables in the parallel coordinate plot to better highlight the differences between species.
2. For the Italian Olive Oils:
   a) Generate a barchart of area and a scatterplot of eicosenoic vs. linoleic. Highlight South-Apulia. How would you describe the marginal distribution in linoleic and eicosenoic acids for the oils from South Apulia relative to the other areas?
   b) Generate a grand tour of oleic, linoleic, arachidic, and eicosenoic. Describe the clustering in the 4D space in relationship to difference between the three regions.
3. For PRIM7:
   a) Sphere the data, that is, transform the data into principal component coordinates.
   b) Run a guided tour using the holes index on the transformed data. Describe the most interesting view found.
4. For Wages, use linking by subject id to identify a man:
   a) whose wage is relatively high after 10 years of experience.
   b) whose wages tended to go down over time.
   c) who had a high salary at 6 years of workforce experience, but much lower salary at 10 years of experience.
   d) who had a high salary at the start of his workforce experience and a lower salary subsequently.
   e) with low wage at the start and a higher salary later.
   f) who has had a relatively high wage, and who has a graduate equivalency diploma.

3
Missing Values

Values are often missing in data, for several reasons. Measuring instruments fail, samples are lost or corrupted, patients do not show up to scheduled appointments, and measurements may be deliberately censored if they are known to be untrustworthy above or below certain thresholds. When this happens, it is always necessary to evaluate the nature and the distribution of the gaps, to see whether a remedy must be applied before further analysis of the data. If too many values are missing, or if gaps on one variable occur in association with other variables, ignoring them may invalidate the results of any analysis that is performed. This sort of association is not at all uncommon and may be directly related to the test conditions of the study. For example, when measuring instruments fail, they often do so under conditions of stress, such as high temperature or humidity. As another example, a lot of the missing values in smoking cessation studies occur for those people who begin smoking again and silently withdraw from the study, perhaps out of discouragement or embarrassment.

In order to prepare the data for further analysis, one remedy that is often applied is imputation, that is, filling in the gaps in the data with suitable replacements. There are numerous methods for imputing missing values. Simple schemes include assigning a fixed value such as the variable mean or median, selecting an existing value at random, or averaging neighboring values. More complex distributional approaches to imputation start with the assumption that the data arises from a standard distribution such as a multivariate normal, which can then be sampled to generate replacement values. See Schafer (1997) for a description of multiple imputation and Little & Rubin (1987) for a description of imputation using multivariate distributions.

In this chapter, we will discuss the power of visual methods at both of these stages: diagnosing the nature and seriousness of the distribution of the missing values (Sect. 3.2) and assessing the results of imputation (Sect. 3.3). The approach follows those described in Swayne & Buja (1998) and Unwin et al. (1996).

3.1 Background

Missing values are classified by Little & Rubin (1987) into three categories according to their dependence structure: missing completely at random (MCAR), missing at random (MAR), and missing not at random (MNAR). Only if values are MCAR is it considered safe to ignore missing values and draw conclusions from the set of complete cases, that is, the cases for which no values are missing. Such missing values may also be called *ignorable*.

The classification of missing values as MCAR means that the probability that a value is missing does not depend on any other observed or unobserved value; that is, $P(missing|observed, unobserved) = P(missing)$. This situation is ideal, where the chance of a value being missing depends on nothing else. It is impossible to verify this classification in practice because its definition includes statements about unobserved data; still, we assume MCAR if there is dependence between missing values and observed data.

In classifying missing values as MAR, we make the more realistic assumption that the probability that a value is missing depends only on the observed variables; that is, $P(missing|observed, unobserved) = P(missing|observed)$. This can be verified with data. For values that are MAR, some structure is allowed as long as all of the structure can be explained by the observed data; in other words, the probability that a value is missing can be defined by conditioning on observed values. This structure of missingness is also called *ignorable*, since conclusions based on likelihood methods are not affected by MAR data.

Finally, when missing values are classified as MNAR, we face a difficult analysis, because $P(missing|observed, unobserved)$ cannot be simplified and cannot be quantified. *Non-ignorable* missing values fall in this category. Here, we have to assume that, even if we condition on all available observed information, the reason for missing values depends on some unseen or unobserved information.

Even when missing values are considered ignorable we may wish to replace them with imputed values, because ignoring them may lead to a non-ignorable loss of data. Consider the following constructed data, where missings are represented by the string NA, meaning "Not Available." There are only 5 missing values out of the 50 numbers in the data:

Case	X_1	X_2	X_3	X_4	X_5
1	NA	20	1.8	6.4	−0.8
2	0.3	NA	1.6	5.3	−0.5
3	0.2	23	1.4	6.0	NA
4	0.5	21	1.5	NA	−0.3
5	0.1	21	NA	6.4	−0.5
6	0.4	22	1.6	5.6	−0.8
7	0.3	19	1.3	5.9	−0.4
8	0.5	20	1.5	6.1	−0.3
9	0.3	22	1.6	6.3	−0.5
10	0.4	21	1.4	5.9	−0.2

Even though only 10% of the numbers in the table are missing, 100% of the variables have missing values, and so do 50% of the cases. A complete case analysis would use only half the data.

Graphical methods can help to determine the appropriate classification of the missing structure as MCAR, MAR or MNAR, and this is described in Sect. 3.2. Section 3.3 describes how the classification can be re-assessed when imputed values are checked for consistency with the data distribution.

3.2 Exploring missingness

One of our first tasks is to explore the distribution of the missing values, seeking to understand the nature of "missingness" in the data. Do the missing values appear to occur randomly, or do we detect a relationship between the missing values on one variable and the recorded values for some other variables in the data? If the distribution of missings is not random, this will weaken our ability to infer structure among the variables of interest. It will be shown later in the chapter that visualization is helpful in searching for the answer to this question.

3.2.1 Shadow matrix

As a first step in exploring the distribution of the missing values, consider the following matrix, which corresponds to the data matrix above and has the same dimensions:

Case	X_1	X_2	X_3	X_4	X_5
1	1	0	0	0	0
2	0	1	0	0	0
3	0	0	0	0	1
4	0	0	0	1	0
5	0	0	1	0	0
6	0	0	0	0	0
7	0	0	0	0	0
8	0	0	0	0	0
9	0	0	0	0	0
10	0	0	0	0	0

In this binary matrix, 1 represents a missing value and 0 a recorded value. It is easier to see the positions of the missing values in this simple version and to consider their distribution apart from the data values. We like to call this the "shadow matrix."

Sometimes there are multiple categories of missingness, in which case this matrix would not simply be binary. For example, suppose we were conducting a longitudinal study in which we asked the same questions of the same subjects over several years. In that case, the missing values matrix might include three values: 0 could indicate an answered question, 1 that a survey respondent failed to answer a question, and 2 that a respondent died before the study was completed.

As an example for working with missing values, we use a small subset of the TAO data: all cases recorded for five locations (latitude 0° with longitudes 110°W and 95°W, 2°S with 110°W and 95°W, and 5°S with 95°W) and two time periods (November to January 1997, an El Niño event, and for comparison, the period from November to January 1993, when conditions were considered normal). Load the data into R and GGobi to investigate missingness:

```
> library(norm)
> library(rggobi)
> d.tao <- read.csv("tao.csv", row.names=1)
> d.tao.93 <- as.matrix(subset(
    d.tao,year==1993,select=sea.surface.temp:vwind))
> d.tao.97 <- as.matrix(subset(
    d.tao,year==1997,select=sea.surface.temp:vwind))
> d.tao.nm.93 <- prelim.norm(d.tao.93)
> d.tao.nm.93$nmis
sea.surface.temp          air.temp         humidity
               3                 4               93
           uwind             vwind
               0                 0
> d.tao.nm.97 <- prelim.norm(d.tao.97)
```

3.2 Exploring missingness   51

```
> d.tao.nm.97$nmis
sea.surface.temp          air.temp         humidity
               0                77                0
           uwind             vwind
               0                 0
```

There are 736 data points, and we find missing values on three of the five variables (Table 3.1).

Table 3.1. Missing values on each variable

Variable	Number of missing values	
	1993	1997
sea surface temp	3	0
air temp	4	77
humidity	93	0
uwind	0	0
vwind	0	0

We are also interested in tabulating missings by case:

```
> d.tao.nm.93$r
    [,1] [,2] [,3] [,4] [,5]
274    1    1    1    1    1
  1    0    0    1    1    1
 90    1    1    0    1    1
  1    1    0    0    1    1
  2    0    0    0    1    1
> d.tao.nm.97$r
    [,1] [,2] [,3] [,4] [,5]
291    1    1    1    1    1
 77    1    0    1    1    1
```

From Table 3.2, we can see that most cases have no missing values (74.5% in 1993, 79.1% in 1997), and less than a quarter of cases have one missing value (24.5% in 1993, 20.9% in 1997). In 1993 two cases are missing two values and two cases have missing values on three of the five variables, sea surface temp, air temp and humidity.

To study the missingness graphically, load the data and set up the colors and glyphs to reflect the year, which corresponds to two different climate conditions.

```
> gd <- ggobi(d.tao)[1]
> glyph_color(gd) <- ifelse(gd[,1]==1993,5,1)
> glyph_type(gd) <- ifelse(gd[,1]==1993,6,4)
```

Table 3.2. Distribution of the number of missing values on a case.

No. of missings on a case	1993 No. of cases	%	1997 No. of cases	%
3	2	0.5	0	0
2	2	0.5	0	0
1	90	24.5	77	20.9
0	274	74.5	291	79.1

3.2.2 Getting started: missings in the "margins"

The simplest approach to drawing scatterplots of variables with missing values is to assign to the missings some fixed value outside the range of the data, and then to draw them as ordinary data points at this unusual location. It is a bit like drawing them in the margins, which is an approach favored in other visualization software. In Fig. 3.1, the three variables with missing values are shown. The missings have been replaced with a value 10% lower than the minimum data value for each variable. In each plot, missing values in the horizontal or vertical variable are represented as points lying along a vertical or horizontal line, respectively. A point that is missing on both variables appears as a point in the lower left corner; if multiple points are missing on both, this point is simply over-plotted.

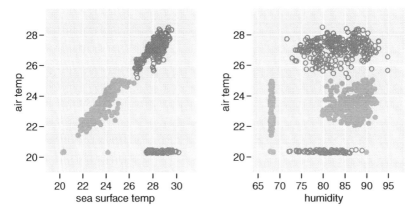

Fig. 3.1. Assigning constants to missing values. In this pair of scatterplots, we have assigned to each missing value a fixed value 10% below the variable minimum, so the "missings" fall along vertical and horizontal lines to the left and below the point scatter. The green solid circles (the cluster that has lower values of air temp) represent data recorded in 1993; the purple open circles show the 1997 data.

What can be seen? Consider the plot of air temp vs. sea surface temp. Not surprisingly, the two temperature values are highly correlated as indicated by the strong linear pattern in the plot; we will make use of this fact a bit later. We can also see that the missings in that plot fall along a horizontal line, telling us that more cases are missing for air temp than for sea surface temp. Some cases are missing for both, and those lie on the point in the lower left corner. The live plot can be queried to find out how many points are over-plotted there. To alleviate the problem of over-plotting, we have also jittered the values slightly; i.e., we have added a small random number to the missing values. In this plot, we also learn that there are no cases missing for sea surface temp but recorded for air temp — if that were true, we would see some points plotted along a vertical line at roughly sea surface temp = 20. The right-hand plot, air temp vs. humidity, is different: There are many cases missing on each variable but not missing on the other.

Both pairwise plots contain the same two clusters of data, one for 1993 records (green filled circles) and the other for 1997, an El Niño year (purple open circles). There is a relationship between the variables and the distribution of the missing values, as we can tell simply from the color of the missings. For example, all cases for which humidity was missing are green, so we know they were all recorded in 1993. The position of the missings on humidity tells the same story, because none of them lie within the range of air temp in 1997. We know already that, if we excluded these cases from our analysis, the results would be distorted: We would exclude 93 out of 368 measurements for 1993, but none for 1997, and the distribution of humidity is quite different in those two years.

The other time period, in 1997, is not immune from missing values either, because all missings for air temp are in purple.

In summary, from these plots we have learned that there is dependence between the missing values and the observed data values. We will see more dependencies between missings on one variable and recorded values on others as we continue to study the data. At best, the missing values here may be MAR (missing at random).

3.2.3 A limitation

Populating missing values with constants is a useful way to begin, as we have just shown. We can explore the data we have and begin our exploration of the missing values as well, because these simple plots allow us to continue using the entire suite of interactive techniques. Multivariate plots, though, such as the tour and parallel coordinate plots are not amenable to this method.

Using fixed values in a tour causes the missing data to be mapped onto artificial planes in p-space, which obscure each other and the main point cloud. Figure 3.2 shows two tour views of sea surface temp, air temp, and humidity with missings set to 10% below minimum. The missing values appear as clustors in the data space, which might be thought of as lying along three walls of a room with the complete data as a scattercloud within the room.

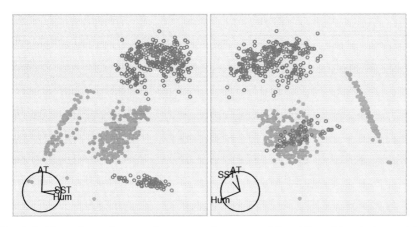

Fig. 3.2. Tour views of sea surface temp, air temp, and humidity with missings set to 10% below minimum. There appear to be four clusters, but two of them are simply the cases that have missings on at least one of the three variables.

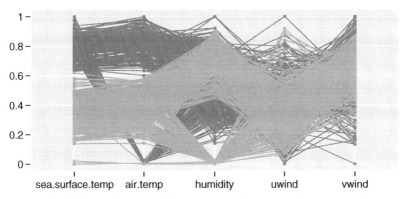

Fig. 3.3. Parallel coordinates of the five variables sea surface temp, air temp, humidity, uwind, and vwind with missings set to 10% below minimum. There are two groups visible for humidity in 1993 (green, the color drawn last), but that is because a large number of missing values are plotted along the zero line; for the same reason, there appear to be two groups for air temp in 1997 (purple).

Figure 3.3 shows the parallel coordinate plot of sea surface temp, air temp, humidity, uwind, and vwind with missings set to 10% below minimum. If we did not know that the points along the zero line were the missings, we could be led to false interpretations of the plot. Consider the values of humidity in 1993 (the green points, the color drawn last), where the large number of points drawn at the zero line look like a second cluster in the data.

When looking at plots of data with missing values, it is important to know whether the missing values have been plotted, and if they have, how they are being encoded.

3.2.4 Tracking missings using the shadow matrix

In order to explore the data and their missing values together, we will treat the shadow matrix (Sect. 3.2.1) as data and display projections of each matrix in linked windows, side by side. In one window, we show the data with missing values replaced by imputed values; in the missing values window, we show the binary indicators of missingness.

Although it may be more natural to display binary data in area plots, we find that scatterplots are often adequate, and we will use them here. We need to spread the points to avoid multiple over-plotting, so we jitter the zeros and ones. The result is a view such as the left-hand plot in Fig. 3.4. The data fall into four squarish clusters, indicating presence and missingness of values for the two selected variables. For instance, the top right cluster consists of the cases for which both variables have missing values, and the lower right cluster shows the cases for which the horizontal variable value is missing but the vertical variable value is present.

Figure 3.4 illustrates the use of the TAO dataset to explore the distribution of missing values for one variable with respect to other variables in the data. We have brushed in orange squares only the cases in the top left cluster, where air temp is missing but humidity is present. We see in the right-hand plot that none of these missings occur for the lowest values of uwind, so we have discovered another dependence between the distribution of missingness on one variable and the distribution of another variable.

We did not really need the missings plot to arrive at this observation; we could have found it just as well by continuing to assign constants to the missing values. In the next section, we will continue to use the missings plot as we begin using imputation.

3.3 Imputation

Although we are not finished with our exploratory analysis of this subset of the TAO data, we have already learned that we need to investigate imputation methods. We have already learned that we will not be satisfied with complete case analysis. We cannot safely throw out all cases with a missing value, because the distribution of the missing values on at least two variables (humidity and air temp) is strongly correlated with at least one other data variable (year).

56   3 Missing Values

Because of this correlation, we need to investigate imputation methods. As we replace the missings with imputed values, though, we do not want to lose track of their locations. We want to use visualization to help us assess imputation methods as we try them, making sure that the imputed values have nearly the same distribution as the rest of the data.

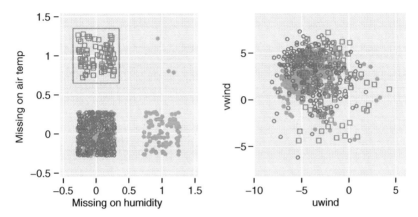

Fig. 3.4. Exploring missingness in the TAO data. The "missings" plot **(left)** for air temp vs. humidity is a jittered scatterplot of zeroes and ones, where one indicates a missing value. The points that are missing only on air temp have been brushed in orange. In a scatterplot of vwind vs. uwind **(right)**, those same missings are highlighted. There are no missings for the very lowest values of uwind.

3.3.1 Mean values

The most rudimentary imputation method is to use the variable mean to fill in the missing values. In the middle row of plots in Fig. 3.5, we have substituted the mean values for missing values on sea surface temp, air temp, and humidity. Even without highlighting the imputed values, some vertical and horizontal lines are visible in the scatterplots. This result is common for any imputation scheme that relies on constants. Another consequence is that the variance–covariance of the data will be reduced, especially if there are a lot of missing values.

3.3.2 Random values

It is clear that a random imputation method is needed to better distribute the replacements. The simplest method is to fill in the missing values with some value selected randomly from among the recorded values for that variable. In the bottom row of plots in Fig. 3.5, we have substituted random values for

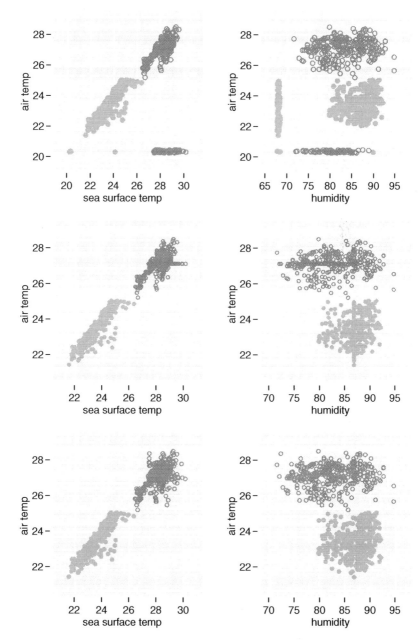

Fig. 3.5. Comparison of simple, widely used imputation schemes. Missings in the margin, as in Fig. 3.1 **(Top row)**. Missing values have been replaced with variable means, conditional on year, producing vertical and horizontal stripes in each cluster **(Middle row)**. Missing values have been filled in by randomly selecting from the recorded values, conditional on year **(Bottom row)**. The imputed values are a little more varied than the recorded data.

missing values on sea surface temp, air temp, and humidity. The match with the data is much better than when mean values were used: It is difficult to distinguish imputed values from recorded data! However, taking random values ignores any association between variables, which results in more variation in values than occurs with the recorded data. If you have a keen eye, you can see that in these plots. It is especially visible in the plot of sea surface temp and air temp, for the 1997 values (in purple): They are more spread, less correlated than the complete data.

The imputed values can be identified using linked brushing between the missings plot and the plot of sea surface temp vs. air temp (Fig. 3.6). Here the values missing on air temp have been brushed (orange rectangles) in the missings plot (left), and we can see the larger spread of the imputed values in the plot at right.

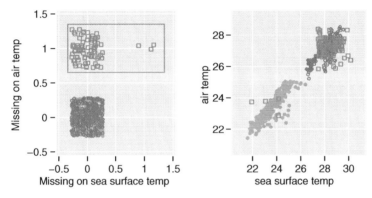

Fig. 3.6. Conditional random imputation. Missing values on all variables have been filled in using random imputation, conditioning on drawing symbol. The imputed values for air temp show less correlation with sea surface temp than do the recorded values.

3.3.3 Multiple imputation

A more sophisticated approach to imputation is to sample from a statistical distribution, which may better reflect the variability in the observed data. Common approaches use regression models or simulation from a multivariate distribution.

To use regression, a linear model is constructed for each of the variables containing missings, with that variable as the response, and the complete data variables as explanatory variables. The distribution of the residuals is used to simulate an error component, which is added to the predicted value for each missing, yielding an imputed value. Many models may be required to impute

3.3 Imputation

all missing values. For example, in the TAO data, we would need to fit a model for sea surface temp, air temp, and humidity separately for each year. This can be laborious!

Simulating from a multivariate distribution yields imputed values from a single model. For the TAO data, it might be appropriate to simulate from a multivariate normal distribution, separately for each year.

With either approach, it is widely acknowledged that one set of imputed values is not enough to measure the variability of the imputation. Multiple imputed values are typically generated for each missing value with these simulation methods, and this process is called *multiple imputation*.

R packages, such as norm by Novo & Schafer (2006) or Hmisc by Dupont & Harrell (2006), contain multiple imputation routines. To view the results in GGobi, we can dynamically load imputed values into a running GGobi process. This next example demonstrates how to impute from a multivariate normal model using R and how to study the results with GGobi. Values are imputed separately for each year.

To apply the light and dark shades used in Fig. 3.7 to the GGobi process launched earlier, select Color Schemes from GGobi's Tools menu and the Paired 10 qualitative color scheme before executing the following lines:

```
> gcolor <- ifelse(gd[,1]==1993,3,9)
> glyph_color(gd) <- gcolor
> ismis <- apply(gd[,4:8], 1, function(x) any(is.na(x)))
> gcolor[ismis] <- gcolor[ismis]+1
> glyph_color(gd) <- gcolor
```

Make a scatterplot of sea surface temp and air temp. At this stage, the missing values are still shown on the line below the minimum of the observed points. In the next step, the missing values are imputed multiply from separate multivariate normal distributions for each of the two years.

```
> rngseed(1234567)
> theta.93 <- em.norm(d.tao.nm.93, showits=TRUE)
Iterations of EM:
1...2...3...4...5...6...7...8...9...10...11...12...13...14...
15...16...17...18...19...20...21...22...23...
> theta.97 <- em.norm(d.tao.nm.97, showits=TRUE)
Iterations of EM:
1...2...3...4...5...6...7...8...9...10...11...12...13...14...
> d.tao.impute.93 <- imp.norm(d.tao.nm.93, theta.93,
      d.tao.93)
> d.tao.impute.97 <- imp.norm(d.tao.nm.97, theta.97,
      d.tao.97)
```

```
> gd[,"sea.surface.temp"] <- c(
    d.tao.impute.97[,"sea.surface.temp"],
    d.tao.impute.93[,"sea.surface.temp"])
> gd[,"air.temp"] = c(
    d.tao.impute.97[,"air.temp"],
    d.tao.impute.93[,"air.temp"])
> gd[,"humidity"] = c(
    d.tao.impute.97[,"humidity"],
    d.tao.impute.93[,"humidity"])
```

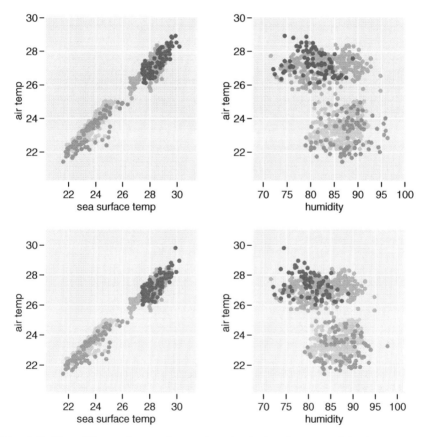

Fig. 3.7. Two different imputations using simulation from a multivariate normal distribution of all missing values. In the scatterplot of air temp vs. sea surface temp the imputed values may have different means than the complete cases: higher sea surface temp and lower air temp. The imputed values of humidity look quite reasonable.

The missings are now imputed, and the scatterplot of sea surface temp and air temp should look like the one in Fig. 3.7. The imputation might make it necessary to re-scale the plot if values have fallen outside the view; if so, use the Rescale button in the Missing Values panel.

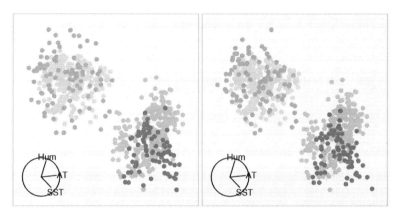

Fig. 3.8. Tour projection of the data after multiple imputation of sea surface temp, air temp, and humidity.

Figures 3.7 and 3.8 show plots of the data containing imputed values resulting from two separate simulations from a multivariate normal mixture. In this coloring, green and purple still mean 1993 and 1997, but now light shades represent recorded values and dark shades highlight the missing values — now imputed.

The imputed values look reasonably good. There are some small differences from the recorded data distribution: Some imputed values for sea surface temp and air temp in 1997 are higher than the observed values, and some imputed values for humidity in 1993 are higher than the observed values.

3.4 Recap

In this chapter, we showed how to use graphical methods to develop a good description of missingness in multivariate data. Using the TAO data, we were able to impute reasonable replacements for the missing values and to use graphics to evaluate them.

The data has two classes, corresponding to two distinct climate patterns, an El Niño event (1997) and a normal season (1993). We discovered the dependence on year as soon we started exploring the missingness, using missings plotted in the margins. Later we discovered other dependencies among the missings and the wind variables using linked brushing between the missings plot (shadow matrix) and other views of the data. These suggest the missing values should be classified as MAR and therefore ignorable, which means that imputation is likely to yield good results.

It was clear that we had to treat these two classes separately in order to get good imputation results, and we imputed values using multiple imputation, simulating from two multivariate normal distributions.

After studying the imputed values, we saw that they were not perfect. Some of the imputed values for air temp and sea surface temp were higher than the observed values. This suggests that the missingness is perhaps MNAR, rather than MAR. Still, the imputed values are close to the recorded values. For practical purposes, it may be acceptable to use them for further analysis of the data.

Exercises

1. Describe the distribution of the wind and temperature variables conditional on the distribution of missing values in humidity, using brushing and the tour.
2. For the Primary Biliary Cirrhosis (PBC) data:
    a) Describe the univariate distributions of complete cases for chol, copper, trig, and platelet. What transformations might be used to make the distributions more bell-shaped? Make these transformations, and use the transformed data for the rest of this exercise.
    b) Examine a scatterplot matrix of chol, copper, trig, platelet with missing values plotted in the margins.
        i. Describe the pairwise relationships among the four variables.
        ii. Describe the distribution between missings and non-missings for trig and platelet.
    c) Generate the shadow matrix, and brush the missing values a different color.
    d) Substitute means for the missing values, and examine the result in a tour. What pattern is obvious among the imputed values?
    e) Substitute random values for the missings, and examine the result in a tour. What pattern is obvious among the imputed values?
    f) In R, generate imputed values using multiple imputation. Examine different sets of imputed values in the scatterplot matrix. Do these sets of values look consistent with the data distribution?
    g) Using spine plots, examine each categorical variable (status, drug, age, sex), checking for associations between the variable and missingness.

4
Supervised Classification

When you browse your email, you can usually tell right away whether a message is spam. Still, you probably do not enjoy spending your time identifying spam and have come to rely on a filter to do that task for you, either deleting the spam automatically or filing it in a different mailbox. An email filter is based on a set of rules applied to each incoming message, tagging it as spam or "ham" (not spam). Such a filter is an example of a supervised classification algorithm. It is formulated by studying a training sample of email messages that have been manually classified as spam or ham. Information in the header and text of each message is converted into a set of numerical variables such as the size of the email, the domain of the sender, or the presence of the word "free." These variables are used to define rules that determine whether an incoming message is spam or ham.

An effective email filter must successfully identify most of the spam without losing legitimate email messages: That is, it needs to be an accurate classification algorithm. The filter must also be efficient so that it does not become a bottleneck in the delivery of mail. Knowing which variables in the training set are useful and using only these helps to relieve the filter of superfluous computations.

Supervised classification forms the core of what we have recently come to call *data mining*. The methods originated in statistics in the early nineteenth century, under the moniker *discriminant analysis*. An increase in the number and size of databases in the late twentieth century has inspired a growing desire to extract knowledge from data, which has contributed to a recent burst of research on new methods, especially on algorithms.

There are now a multitude of ways to build classification rules, each with some common elements. A training sample contains data with known categorical response values for each recorded combination of explanatory variables. The training sample is used to build the rules to predict the response. Accuracy, or inversely error, of the classifier for future data is also estimated from the training sample. Accuracy is of primary importance, but there are many other interesting aspects of supervised classification applications beyond this:

- Are the classes well separated in the data space, so that they correspond to distinct clusters? If so, what are the shapes of the clusters? Is each cluster sufficiently ellipsoidal so that we can assume that the data arises from a mixture of multivariate normal distributions? Do the clusters exhibit characteristics that suggest one algorithm in preference to others?
- Where does the boundary between classes fall? Are the classes linearly separable, or does the difference between classes suggest a non-linear boundary? How do changes in the input parameters affect these boundaries? How do the boundaries generated by different methods vary?
- What cases are misclassified, or have more uncertain predictions? Are there places in the data space where predictions are especially good or bad?
- Is it possible to reduce the set of explanatory variables?

This chapter discusses the use of interactive and dynamic graphics to investigate these different aspects of classification problems. It is structured as follows: Sect. 4.1 gives a brief background of the major approaches, Sect. 4.2 describes graphics for viewing the classes, and Sect. 4.3 adds descriptions of several different numerical methods and graphics to assess the models they generate. A good companion to this chapter is the material presented in Venables & Ripley (2002), which provides data and code for practical examples of supervised classification using R. For the most part, methods described here are well documented in the literature. Consequently, our descriptions are brief, and we provide references to more in-depth explanations. We have gone into more detail at the beginning of the chapter, to set the tone for the rest of the chapter, and in the section on support vector machines at the end of the chapter, because we find the descriptions in the literature to be unsatisfactory.

4.1 Background

Supervised classification arises when there is a categorical response variable (the output) $Y_{n\times 1}$ and multiple explanatory variables (the input) $\mathbf{X}_{n\times p}$, where n is the number of cases in the data and p is the number of variables. Because Y is categorical, the user may represent it using character strings that the software will recode as integers. A binary variable may be converted from $\{Male, Female\}$ or $\{T, F\}$ to $\{1, 0\}$ or $\{-1, 1\}$, whereas multiple classes may be recoded using the values $\{1, \ldots, g\}$. Coding of the response really matters and can alter the formulation or operation of a classifier.

Since supervised classification is used in several disciplines, the terminology used to describe the elements can vary widely. The attributes may also be called features, independent variables, or explanatory variables. The instances may be called cases, rows, or records.

4.1.1 Classical multivariate statistics

Discriminant analysis dates to the early 1900s. Fisher's linear discriminant (Fisher 1936) determines a linear combination of the variables that separates two classes by comparing the differences between class means with the variance of values within each class. It makes no assumptions about the distribution of the data. Linear discriminant analysis (LDA), as proposed by Rao (1948), formalizes Fisher's approach by imposing the assumption that the data values for each class arise from a p-dimensional multivariate normal distribution, which shares a common variance–covariance matrix with data from other classes. Under this assumption, Fisher's linear discriminant gives the optimal separation between the two groups.

For two equally weighted groups, where Y is coded as $\{0, 1\}$, the LDA rule is:

Allocate a new observation $\mathbf{X}_0$ to group 1 if

$$(\bar{\mathbf{X}}_1 - \bar{\mathbf{X}}_2)' \mathbf{S}_{\text{pooled}}^{-1} \mathbf{X}_0 \geq \frac{1}{2}(\bar{\mathbf{X}}_1 - \bar{\mathbf{X}}_2)' \mathbf{S}_{\text{pooled}}^{-1}(\bar{\mathbf{X}}_1 + \bar{\mathbf{X}}_2)$$

else allocate it to group 2,

where $\bar{\mathbf{X}}_k$ are the class mean vectors of an $n \times p$ data matrix $\mathbf{X}_k$ $(k = 1, 2)$,

$$\mathbf{S}_{\text{pooled}} = \frac{(n_1 - 1)\mathbf{S}_1}{(n_1 - 1) + (n_2 - 1)} + \frac{(n_2 - 1)\mathbf{S}_2}{(n_1 - 1) + (n_2 - 1)}$$

is the pooled variance–covariance matrix, and

$$\mathbf{S}_k = \frac{1}{n-1} \sum_{i=1}^{n} (\mathbf{X}_{ki} - \bar{\mathbf{X}}_k)(\mathbf{X}_{ki} - \bar{\mathbf{X}}_k)', \quad k = 1, 2$$

is the class variance–covariance matrix. The linear discriminant part of this rule is $(\bar{\mathbf{X}}_1 - \bar{\mathbf{X}}_2)' \mathbf{S}_{\text{pooled}}^{-1}$, which defines the linear combination of variables that best separates the two groups. To define a classification rule, we compute the value of the new observation $\mathbf{X}_0$ on this line and compare it with the value of the average of the two class means $(\bar{\mathbf{X}}_1 + \bar{\mathbf{X}}_2)/2$ on the same line.

For multiple (g) classes, the rule and the discriminant space are constructed using the between-group sum-of-squares matrix,

$$\mathbf{B} = \sum_{k=1}^{g} n_k (\bar{\mathbf{X}}_k - \bar{\mathbf{X}})(\bar{\mathbf{X}}_k - \bar{\mathbf{X}})'$$

which measures the differences between the class means, compared with the overall data mean $\bar{\mathbf{X}}$ and the within-group sum-of-squares matrix,

$$\mathbf{W} = \sum_{k=1}^{g} \sum_{i=1}^{n_k} (\mathbf{X}_{ki} - \bar{\mathbf{X}}_k)(\mathbf{X}_{ki} - \bar{\mathbf{X}}_k)'$$

66   4 Supervised Classification

which measures the variation of values around each class mean. The linear discriminant space is generated by computing the eigenvectors (canonical coordinates) of $\mathbf{W}^{-1}\mathbf{B}$, and this is the space where the group means are most separated with respect to the pooled variance–covariance. The resulting classification rule is to allocate a new observation to the class with the highest value of

$$\bar{\mathbf{X}}'_k \mathbf{S}^{-1}_{\text{pooled}} \mathbf{X}_0 - \frac{1}{2}\bar{\mathbf{X}}'_k \mathbf{S}^{-1}_{\text{pooled}} \bar{\mathbf{X}}_k \quad k=1,...,g \qquad (4.1)$$

which results in allocating the new observation into the class with the closest mean.

This LDA approach is widely applicable, but it is useful to check the underlying assumptions on which it depends: (1) that the cluster structure corresponding to each class forms an ellipse, showing that the class is consistent with a sample from a multivariate normal distribution, and (2) that the variance of values around each mean is nearly the same. Figure 4.1 illustrates two datasets, of which only one is consistent with these assumptions. Other parametric models, such as quadratic discriminant analysis or logistic regression, also depend on assumptions about the data which should be validated.

Our description is derived from Venables & Ripley (2002) and Ripley (1996). A good general treatment of parametric methods for supervised classification can be found in Johnson & Wichern (2002) or another similar multivariate analysis textbook. Missing from multivariate textbooks is a good explanation of the use of interactive graphics both to check the assumptions underlying the methods and to explore the results. This chapter fills this gap.

4.1.2 Data mining

Algorithmic methods have overtaken parametric methods in the practice of supervised classification. A parametric method such as linear discriminant analysis yields a set of interpretable output parameters, so it leaves a clear trail helping us to understand what was done to produce the results. An algorithmic method, on the other hand, is more or less a black box, with various input parameters that are adjusted to tune the algorithm. The algorithm's input and output parameters do not always correspond in any obvious way to the interpretation of the results. All the same, these methods can be very powerful and their use is not limited by requirements about variable distributions as is the case with parametric methods.

The tree algorithm (Breiman, Friedman, Olshen & Stone 1984) is a widely used algorithmic method. The tree algorithm generates a classification rule by sequentially splitting the data into two buckets. Splits are made between sorted data values of individual variables, with the goal of obtaining pure

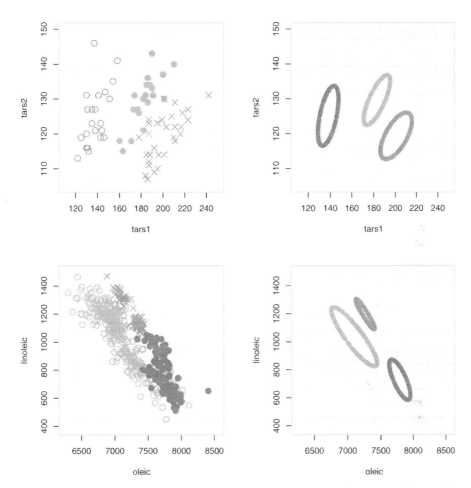

Fig. 4.1. Evaluating model assumptions by comparing scatterplots of raw data with bivariate normal variance–covariance ellipses. For the Flea Beetles (**top row**), each of the three classes in the raw data (**left**) appears consistent with a sample from a bivariate normal distribution with equal variance–covariance. For the Olive Oils, the clusters are not elliptical, and the variance differs from cluster to cluster.

classes on each side of the split. The inputs for a simple tree classifier commonly include (1) an impurity measure, an indication of the relative diversity among the cases in the terminal nodes; (2) a parameter that sets the minimum number of cases in a node, or the minimum number of observations in a terminal node of the tree; and (3) a complexity measure that controls the growth of a tree, balancing the use of a simple generalizable tree against a more accurate tree tailored to the sample. When applying tree methods, exploring the effects of the input parameters on the tree is instructive; for example, it helps us to assess the stability of the tree model.

Although algorithmic models do not depend on distributional assumptions, that does not mean that every algorithm is suitable for all data. For example, the tree model works best when all variables are independent within each class, because it does not take such dependencies into account. As always, visualization can help us to determine whether a particular model should be applied. In classification problems, it is useful to explore the cluster structure, comparing the clusters with the classes and looking for evidence of correlation within each class. The upper left-hand plot in Fig. 4.1 shows a strong correlation between tars1 and tars2 within each cluster, which indicates that the tree model may not give good results for the Flea Beetles. The plots in Fig. 4.2 provide added evidence. They use background color to display the class predictions for LDA and a tree. The LDA boundaries, which are formed from a linear combination of tars1 and tars2, look more appropriate than the rectangular boundaries of the tree classifier.

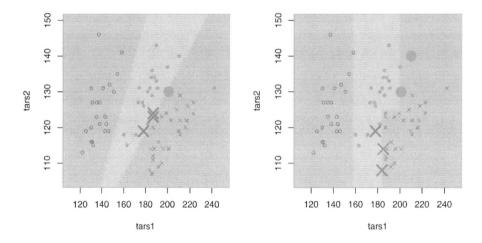

Fig. 4.2. Classification of the data space for the Flea Beetles, as determined by LDA **(left)** and a tree model **(right)**. Misclassified cases are highlighted.

Hastie, Tibshirani & Friedman (2001) and Bishop (2006) include thorough discussions of algorithms for supervised classification presented from a modeling perspective with a theoretical emphasis. Ripley (1996) is an early volume describing and illustrating both classical statistical methods and algorithms for supervised classification. All three books contain some excellent examples of the use of graphics to examine two-dimensional (2D) boundaries generated by different classifiers. The discussions in these and other writings on data mining algorithms take a less exploratory approach than that of this chapter, and they lack treatments of the use of graphics to examine the high-dimensional spaces in which the classifiers operate.

4.1.3 Studying the fit

A classifier's performance is usually assessed using its error or, conversely, its accuracy. Error is calculated by comparing the predicted class with the known true class, using a misclassification table. For example, below are the respective misclassification tables for LDA and the tree classifier applied to the Flea Beetles:

LDA

		Predicted Class			Error
		1	2	3	
	1	20	0	1	0.048
Class	2	0	22	0	0.000
	3	3	0	28	0.097
					0.054

Tree

		Predicted Class			Error
		1	2	3	
	1	19	0	2	0.095
Class	2	0	22	0	0.000
	3	3	0	28	0.097
					0.068

The total error is the number of misclassified samples divided by the total number of cases: $4/74 = 0.054$ for LDA and $5/74 = 0.068$ for the tree classifier.

It is informative to study the misclassified cases and to see which pockets of the data space contain more error. The misclassified cases for LDA and tree classifiers are highlighted (large orange ×es and large green circles) in Fig. 4.2. Some errors made by the tree classifier, such as the uppermost large green circle, seem especially egregious. As noted earlier, they result from the limitations of the algorithm when variables are correlated.

To be useful, the error estimate should predict the performance of the classifier on new samples not yet seen. However, if the error is calculated using the same data that was used by the classifier, it is likely to be too low. Many methods are used to avoid double-dipping from the data, including several types of *cross-validation*. A simple example of cross-validation is to split the data into a training sample (used by the classifier) and a test sample (used for calculating error).

Ensemble methods build cross-validation into the error calculations. Ensembles are constructed by using multiple classifiers and by pooling the predictions using a voting scheme. A random forest (Breiman 2001, Breiman & Cutler 2004), for example, builds in cross-validation by constructing multiple trees, each of which is generated by randomly sampling the input variables and the cases. Because each tree is built using a sample of the cases, there is in effect a training sample and a test sample for each tree. (See Sect. 4.3.3 for more detail.)

4.2 Purely graphics: getting a picture of the class structure

To visualize the class structure, we start by coding the response variable Y using color and symbol to represent class, and then we explore a variety of plots of the explanatory variables $\mathbf{X}$. Our objective is to learn how distinctions between classes arise. If we are lucky, we will find views in which there are gaps between clusters corresponding to the classes. A gap indicates a well-defined distinction between classes and suggests that there will be less error in predicting future samples. We will also study the shape of the clusters.

If the number of classes is large, keep in mind that it is difficult to digest information from plots having more than three or four colors or symbols. You may be able to simplify the displays by grouping classes into a smaller set of "super-classes." Alternatively, you can partition the data, looking at a few classes at a time.

If the number of dimensions is large, it takes much longer to get a sense of the data, and it is easy to get lost in high-dimensional plots. There are many possible low-dimensional plots to examine, and that is the place to start. Explore plots one or two variables at a time before building up to multivariate plots.

4.2.1 Overview of Italian Olive Oils

The Olive Oils data has eight explanatory variables (levels of fatty acids in the oils) and nine classes (areas of Italy). The goal of the analysis is to develop rules that reliably distinguish oils from the nine different areas. It is a problem of practical interest, because oil from some areas is more highly valued and unscrupulous suppliers sometimes make false claims about the origin of their oil. The content of the oils is a subject of study in its own right: Olive oil has high nutritional value, and some of its constituent fatty acids are considered to be more beneficial than others.

In addition, this is a good dataset for students of supervised classification because it contains a mix of straightforward separations, difficult separations, and unexpected finds.

As suggested above, we do not start by trying to visualize or classify the oils by area, because nine groups are too many. Instead, we divide the classification job into a two-stage process. We start by grouping the nine areas into three "super-classes" corresponding to a division of Italy into South, North, and Sardinia, and we call this new variable region. In the first stage, we classify the oils by region into three groups. In the second stage, we work with the oils from one region at a time, building classifiers to predict area within region.

4.2 Purely graphics: getting a picture of the class structure

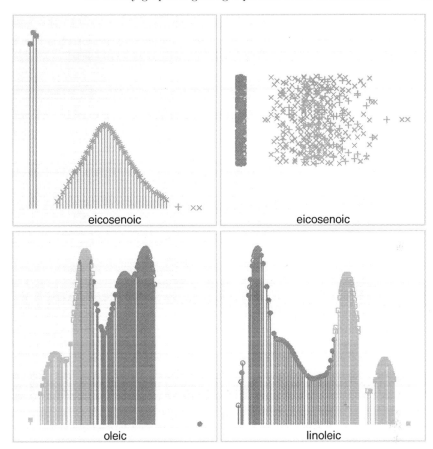

Fig. 4.3. Differentiating the oils from the three regions in Olive Oils in univariate plots. **(top row)** eicosenoic separates Southern oils (in orange ×es and +es) from the others, as shown in both an ASH and a textured dot plot. **(bottom row)** In plots where Southern oils have been removed, we see that Northern (purple circles) and Sardinian (green rectangles) oils are separated by linoleic, although there is no gap between the two clusters.

4.2.2 Building classifiers to predict region

Univariate plots: We first paint the points according to region. Using univariate plots, we look at each explanatory variable in turn, looking for separations between pairs of regions. This table describes the correspondence between region and symbol for the next few figures:

region	Symbol
South	orange + and ×
Sardinia	green rectangle
North	purple circle

4 Supervised Classification

We can cleanly separate the oils of the South from those of the other regions using just one variable, eicosenoic (Fig. 4.3, top row). Both of these univariate plots show that the oils from the other two regions contain no eicosenoic acid.

In order to differentiate the oils from the North and Sardinia, we remove the Southern oils from view and continue plotting one variable at a time (Fig. 4.3, bottom row). Several variables show differences between the oils of the two regions, and we have plotted two of them: oleic and linoleic. Oils from Sardinia contain lower amounts of oleic acid and higher amounts of linoleic acid than oils from the north. The two regions are perfectly separated by linoleic, but since there is no gap between the two groups of points, we will keep looking.

Bivariate plots: If one variable is not enough to distinguish Northern oils from Sardinian oils, perhaps we can find a pair of variables that will do the job. Starting with oleic and linoleic, which were so promising when taken singly, we look at pairwise scatterplots (Fig. 4.4, left and middle). Unfortunately, the combination of oleic and linoleic is no more powerful than each one was alone. They are strongly negatively associated, and there is still no gap between the two groups.

We explore other pairs of variables. Something interesting emerges from a plot of arachidic and linoleic: There is big gap between the points of the two regions! Arachidic alone seems to have no power to separate, but it improves the power of linoleic. Since the gap between the two groups follows a non-linear, almost quadratic path, we must do a bit more work to define a functional boundary.

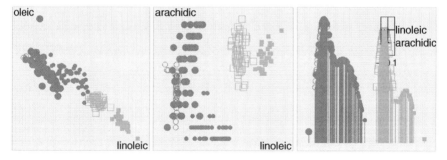

Fig. 4.4. Separation between the Northern (purple circles) and Sardinian (green squares) oils. Two bivariate scatterplots **(left)** and a linear combination of linoleic and arachidic viewed in a 1D tour **(right)**.

We move on, using the 1D tour to look for a linear combination of linoleic and arachidic that will show a clear gap between the Northern and Sardinian oils, and we find one (Fig. 4.4, right). The linear combination is

4.2 Purely graphics: getting a picture of the class structure

composed mostly of linoleic with a small contribution from arachidic. (The numbers generating this projection, recorded from the tour coefficients, are $\frac{0.969}{1022} \times$ linoleic $+ \frac{0.245}{105} \times$ arachidic. The numerator is the projection coefficient, and the denominator is the range of the variable, which was used to scale the measurements into the plot.) We usually think of this as a multivariate plot, but here we were successful using a linear combination of only two variables.

Multivariate plots: A parallel coordinate plot can also be used to select important variables for classification. Figure 4.5 shows a parallel coordinate plot for the Olive Oils data, where the three colors represent the three large regions. As we found earlier, eicosenoic is useful for separating Southern oils (orange, the color drawn first) from the others. In addition, Southern oils have higher values on palmitic and palmitoleic and low values on oleic. Northern oils have high values of oleic and low values of linoleic relative to Sardinian oils.

Parallel coordinate plots are not as good as tours for visualizing the shape of the clusters corresponding to classes and the shape of the boundaries between them, but they are attractive because they can hold so many variables at once and still be clearly labeled.

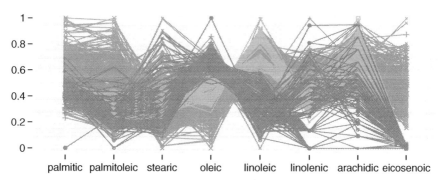

Fig. 4.5. Finding variables for classification using a parallel coordinate plot of Olive Oils. Color represents region, with South in orange (the color that is drawn first), North in purple (the color that is drawn last), and Sardinia in green. Eicosenoic, and to some extent palmitic, palmitoleic and oleic distinguish Southern oils from the others. Oleic and linoleic distinguish Northern from Sardinian oils.

4.2.3 Separating the oils by area within each region

It is now clear that the oils from the three large regions can be distinguished by their fatty acid composition. For the second stage of the classification task, we explore one region at a time, looking for separations among the oils of each area. We plan to separate them visually just as we did in the preceding section,

74   4 Supervised Classification

by starting with univariate plots and adding dimensions using bivariate and multivariate plots if necessary.

Northern Italy: The region "North" was created by aggregating three areas: Umbria, East Liguria, and West Liguria. This table describes the correspondence between area and symbol for the next figure:

area	Symbol
Umbria	pink open circles
East Liguria	purple solid circles
West Liguria	blue solid circles

The univariate plots show no clear separations of oils by area, although several variables are correlated with area. For example, oils from West Liguria have higher linoleic acid content than those from other two areas (Fig. 4.6, top left). The bivariate plots, too, fail to show clear separations by area, but two variables, stearic and linoleic, look useful (Fig. 4.6, top right). Oils from West Liguria have the highest linoleic and stearic acid content, and oils from Umbria have the lowest linoleic and stearic acid content.

Starting with these two variables, we explore linear combinations using a 1D tour, looking for other variables that further separate the oils by area. We used projection pursuit guidance, with the LDA index, to find the linear combination shown in Fig. 4.6 (bottom left). West Liguria is almost separable from the other two areas using a combination of palmitoleic, stearic, linoleic, and arachidic.

At this point, we could proceed in two different directions. We could remove the points corresponding to West Ligurian oils and look for differences between the other two areas of the Northern region, or we could use 2D projections to look for a better separation. We choose the latter, and find a 2D linear combination of the same four variables (palmitoleic, stearic, linoleic, and arachidic), which almost separates the oils from the three areas (Fig. 4.6, bottom right). We found it using projection pursuit guidance with the LDA index, followed by some manual tuning.

There are certainly clusters corresponding to the three areas, but we have not found a projection in which they do not overlap. It may not be possible to build a classifier that perfectly predicts the areas within the region North, but the error should be very small.

Sardinia: This region is composed of two areas, Coastal and Inland Sardinia. This is an easy one! We leave it to the reader to look at a scatterplot of oleic and linoleic, which shows a big gap between two clusters corresponding to the two areas.

Southern Italy: In this data, four areas are grouped into the region "South." These are North Apulia, South Apulia, Calabria, and Sicily. This table describes the correspondence between area and symbol for the next figure:

4.2 Purely graphics: getting a picture of the class structure   75

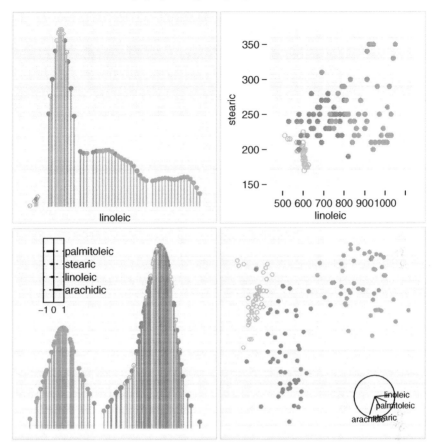

Fig. 4.6. Separating the oils of the Northern region by area. The 1D ASH **(top left)** shows that oils from West Liguria (the blue cluster at the right of the plot) have a higher percentage of linoleic. Looking for other variables, we see in the bivariate scatterplot **(top right)** that stearic and linoleic almost separate the three areas. The 1D and 2D tour plots **bottom row** show that linear combinations of palmitoleic, stearic, linoleic, and arachidic are useful for classification.

area	Symbol
North Apulia	orange +
Calabria	red +
South Apulia	pink ×
Sicily	yellow ×

The prospect of finding visual separations between these four areas looks dismal. In a scatterplot of palmitoleic and palmitic (Fig. 4.7, top row), there is a big gap between North Apulia (low on both variables) and South Apulia (high on both variables), with Calabria in the middle. The troublemaker is

Sicily: The cluster of oils from Sicily overlaps those of the other three areas. (The plots in both rows are the same, except that the Sicilian oils have been excluded from the bottom row of plots.)

This pattern does not change much when more variables are used. We can separate oils from Calabria, North Apulia, and South Apulia pretty well in a 2D tour, but oils from Sicily overlap the oils from every other area in every projection. The plot at the top right of Fig. 4.7 is a typical example; it was found using projection pursuit with the LDA index and then adjusted using manual controls.

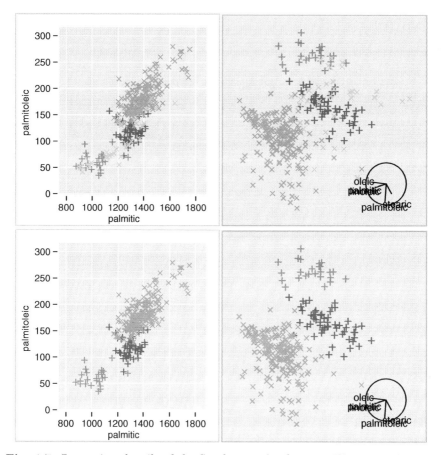

Fig. 4.7. Separating the oils of the Southern region by area. 2D tour projections including all variables (**right**) show that the oils of Southern Italy are almost separable by area. The samples from Sicily (yellow ×), however, overlap the points of the other three areas, as can be seen by comparing the plots in the top row with those in the bottom row, from which Sicilian oils have been excluded.

4.2.4 Taking stock

The Olive Oils have dramatically different fatty acid composition depending on geographic region. The three geographic regions, created by aggregating the areas into North, South and Sardinia, are well separated based on eicosenoic, linoleic, and arachidic.

The oils from the North are mostly separable from each other by area, using all variables. The oils from the inland and coastal regions of Sardinia have different amounts of oleic and linoleic acids.

The oils from three areas in the South are almost separable, but the oils from Sicily can not be separated. Why are these oils indistinguishable from the oils of the other areas in the South? Is there a problem with the quality of these samples?

4.3 Numerical methods

In this section, we show how classification algorithms can be supported and enhanced by graphical methods. Graphics should be used before modeling, to make sure the data conforms to the assumptions of the model, and they are equally useful after modeling, to assess the fit of the model, study the failures, and compare the results of different models. Without graphics, it is easy to apply an algorithm inappropriately and to achieve results that seem convincing but have little or no meaning.

We will discuss graphical methods for linear discriminant analysis, trees, random forests, neural networks, and support vector machines, as applied to the Olive Oils. References are provided for the reader who wants to know more about the algorithms, which are only briefly described here.

4.3.1 Linear discriminant analysis

LDA, which has already been discussed extensively in this chapter, is used to find the linear combination of variables that best separates classes. When the variance–covariance structure of each class is (1) ellipsoidal and (2) roughly equal, LDA may be used to build a classifier. Even when those assumptions are violated, the linear discriminant space may provide a good low-dimensional view of the cluster structure. This low-dimensional view can then be used by other classifiers with less stringent requirements.

To determine whether it is safe to use LDA to classify a particular dataset, we can use the 2D tour to test the assumptions made by the model. We view a large number of projections of the data, and we want to see ellipses of roughly the same size in all projections. Turning back to look at Fig. 4.1, we are reminded that LDA is not an appropriate classifier for the Olive Oils, because the variance–covariances are neither equal nor elliptical for all classes.

78   4 Supervised Classification

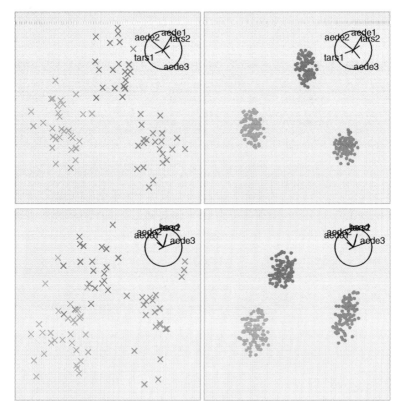

Fig. 4.8. Evaluating model assumptions by comparing scatterplots of the Flea Beetles data **(left)** with corresponding projections of 6D normal variance–covariance ellipsoids **(right)**.

Earlier in the chapter, LDA seemed suitable for another of our datasets, so we detour briefly from our examination of Olive Oils to consider Flea Beetles again. We have already seen that the first two variables of the Flea Beetles data (Fig. 4.1) are consistent with the equal variance–covariance, multivariate normal model, but before using LDA we would need to be sure that the assumptions hold for all six variables. We rotated six variables in the 2D tour for a while, and Fig. 4.8 shows two of the many projections we observed. In this figure, we compare the projection of the data (at left) with the corresponding projection of the 6D variance–covariance ellipsoids (at right). In some projections (bottom row), there are a few slight deviations from the equal ellipsoidal structure, but the differences are small enough to be due to sampling variability. We are satisfied that it would be appropriate to apply LDA to the Flea Beetle data.

Even when the LDA model assumptions are violated, we can still use it to find a good low-dimensional view of the cluster structure (as discussed in

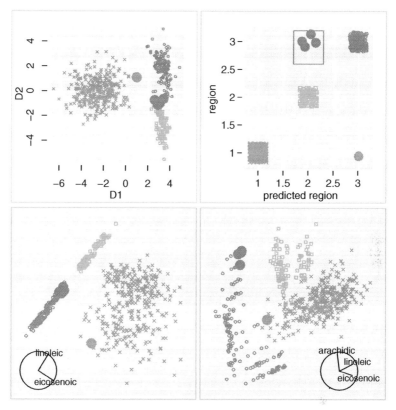

Fig. 4.9. Misclassifications from an LDA classifier of the Olive oils by region. In the misclassification table **(top right)**, erroneously classified samples are brushed as large filled circles, and studied in the discriminant space **(top left)**, and the 2D tour **(bottom row)**.

Sect. 4.1.1). This is the discriminant space, the linear combination of variables where the class means are most separated relative to the pooled variance–covariance. For the Olive Oils, it is computed by:

```
> library(MASS)
> library(rggobi)
> d.olive <- read.csv("olive.csv", row.names=1)
> d.olive.sub <- subset(d.olive,
    select=c(region,palmitic:eicosenoic))
> olive.lda <- lda(region~., d.olive.sub)
> pregion <- predict(olive.lda, d.olive.sub)$class
```

80   4 Supervised Classification

```
> table(d.olive.sub[,1], pregion)
   pregion
      1   2   3
 1  322   0   1
 2    0  98   0
 3    0   4 147
> plot(predict(olive.lda, d.olive.sub)$x)
> gd <- ggobi(cbind(d.olive, pregion))[1]
> glyph_color(gd) <- c(rep(6,323), rep(5,98), rep(1,151))
```

The top left plot in Fig. 4.9 shows data projected into the discriminant space for the Olive oils. Oils from the South form the large well-separated cluster at the left; oils from the North are at the top right of the plot, and are not well separated from the Sardinian oils just below them.

The misclassifications made by the model are highlighted in the plot, drawn with large filled circles, and we can learn more about LDA by exploring them:

		Predicted region			Error
		South	Sardinia	North	
	South	322	0	1	0.003
region	Sardinia	0	98	0	0.000
	North	0	4	147	0.026
					0.009

If we use LDA as a classifier, five samples are misclassified. It is not at all surprising to see misclassifications where clusters overlap, as the Northern and Sardinian regions do, so the misclassification of four Northern samples as Sardinian is not troubling.

One very surprising misclassification is represented by the orange circle, showing that, despite the large gap between these clusters, one of the oils from the South has been misclassified as a Northern oil. As discussed earlier, LDA is blind to the size of the gap when its assumptions are violated. Since the variance–covariance of these clusters is so different, LDA makes obvious mistakes, placing the boundary too close to the Southern oils, the group with largest variance.

These misclassified samples are examined in other projections shown by a tour (bottom row of plots). The Southern oil sample that is misclassified is on the outer edge of the cluster of oils from the South, but it is very far from the points from the other regions. It really should not be confused — it is clearly a Southern oil. Actually, even the four misclassified samples from the North should not be confused by a good classifier, because even though they are at one edge of the cluster of Northern oils, they are still far from the cluster of Sardinian oils.

As we mentioned above, even when LDA fails as a classifier, the discriminant space can be a good starting place to manually search the neighborhood for a clearer view, that is, to sharpen the image. It can usually be re-created in a projection pursuit guided tour using the LDA index. The projection shown in the lower left-hand plot of Fig. 4.9 in which the three regions, especially North and Sardinia, are better separated, was found using manual controls starting from a local maximum of the LDA index.

4.3.2 Trees

Tree classifiers generally provide a simpler solution than linear discriminant analysis. For example, on the Olive Oils, the tree formed here:

```
> library(rpart)
> olive.rp <- rpart(region~., d.olive.sub, method="class")
> olive.rp
```

yields this solution:

if eicosenoic $\geq$ 6.5 assign the sample to South
else
 if linoleic $\geq$ 1053.5 assign the sample to Sardinia
 else assign the sample to North

(There may be very slight variation in solutions depending on the tree implementation and input to the algorithm.) This rule is simple because it uses only two of the eight variables, and it is also accurate, yielding no misclassifications; that is, it has zero prediction error.

The tree classifier is constructed by an algorithm that examines the values of each variable for locations where a split would separate members of different classes; the best split among the choices in all the variables is chosen. The data is then divided into the two subsets falling on each side of the split, and the process is then repeated for each subset, until it is not prudent to make further splits.

The numerical criteria of accuracy and simplicity suggest that the tree classifier for the Olive Oil is perfect. However, by examining the classification boundary plotted on the data (Fig. 4.10, top left), we see that it is not: Oils from Sardinia and the North are not clearly differentiated. The tree method does not consider the variance–covariance of the groups — it simply slices the data on single variables. The separation between the Southern oils and the others is wide, so the algorithm finds that first, and slices the data right in the middle of the gap. It next carves the data between the Northern and Sardinian oils along the linoleic axis — even though there is no gap between these groups along that axis. With so little separation between these two classes, the solution may be quite unstable for future samples of Northern and Sardinian oils: A small difference in linoleic acid content may cause a new observation to be assigned into the wrong region.

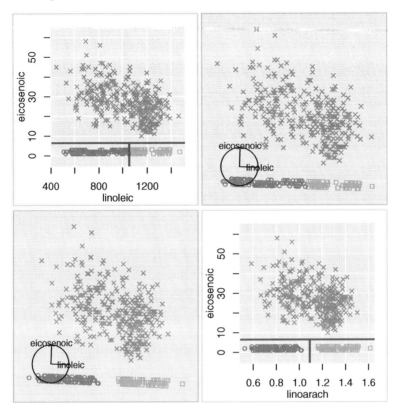

Fig. 4.10. Improving on the results of the tree classifier using the manual tour. The tree classifier determines that only eicosenoic and linoleic acid are necessary to separate the three regions (**R plot, top left**). This view is duplicated in GGobi (**top right**) and sharpened using manual controls (**bottom left**). The improved result is then returned to R and re-plotted with reconstructed boundaries (**bottom right**).

Tree classifiers are usually effective at singling out the most important variables for classification. However, since they define each split using a single variable, they are likely to miss any model improvements that might come from using linear combinations of variables. Some tree implementations consider linear combinations of variables, but they are not in common use. The more commonly used models might, at best, approximate a linear combination by using many splits along the different variables, zig-zagging a boundary between clusters.

Accordingly the model produced by a tree classifier can sometimes be improved by exploring the neighborhood using the manual tour controls (Fig. 4.10, top right and bottom left). Starting from the projection of the two variables selected by the tree algorithm, linoleic and eicosenoic, we find an improved projection by including just one other variable, arachidic. The gap between the Northern and Sardinian regions is distinctly wider.

We can capture the coefficients of rotation that generate this projection, and we can use them to create a new variable. We define *linoarach* to be $\frac{0.969}{1022} \times$ linoleic $+ \frac{0.245}{105} \times$ arachidic. We can add this new variable to the olive oils data and run the tree classifier on the augmented data. The new tree is:

> *if eicosenoic* $>= 6.5$ *assign the sample to South*
> *else*
> > *if linoarach* $>= 1.09$ *assign the sample to Sardinia*
> > *else assign the sample to North*

Is this tree better than the original? The error for both models is zero, so there is no difference numerically. However, the plots in Fig. 4.10 (top left and bottom right) suggest that the sharpened tree would be more robust to small variations in the fatty acid content and would thus classify new samples with less error.

4.3.3 Random forests

A random forest (Breiman 2001, Breiman & Cutler 2004) is a classifier that is built from multiple trees generated by randomly sampling the cases and the variables. The random sampling (with replacement) of cases has the fortunate effect of creating a training ("in-bag") and a test ("out-of-bag") sample for each tree computed. The class of each case in the out-of-bag sample for each tree is predicted, and the predictions for all trees are combined into a vote for the class identity.

A random forest is a computationally intensive method, a "black box" classifier, but it produces various diagnostics that make the outcome less mysterious. Some diagnostics that help us to assess the model are the votes, the measures of variable importance, the error estimate, and as usual, the misclassification tables.

We test the method on the Olive Oils by building a random forest classifier of 500 trees, using the R package randomForest (Liaw, Wiener, Breiman & Cutler 2006):

```
> library(randomForest)
> olive.rf <- randomForest(as.factor(region)~.,
    data=d.olive.sub, importance=TRUE, proximity=TRUE, mtry=4)
> order(olive.rf$importance[,5], decreasing=T)
[1] 8 5 4 1 7 2 6 3
> pred <- as.numeric(olive.rf$predicted)
```

84   4 Supervised Classification

```
> table(d.olive.sub[,1], olive.rf$predicted)
       1   2   3
  1  323   0   0
  2    0  98   0
  3    0   0 151
> margin <- olive.rf$vote
> colnames(margin) <- c("Vote1", "Vote2", "Vote3")
> d.olive.rf <- cbind(pred, margin, d.olive)
> gd <- ggobi(d.olive.rf)[1]
> glyph_color(gd) <- c(rep(6,323), rep(5,98), rep(1,151))
```

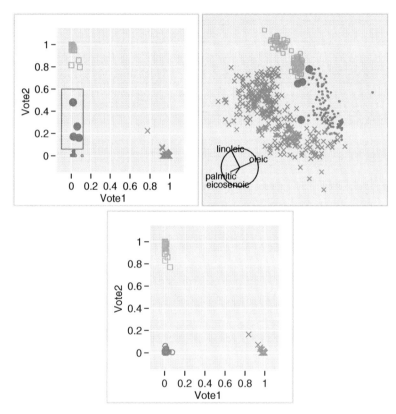

Fig. 4.11. Examining the results of a forest classifier of Olive Oils by region. The votes assess the uncertainty associated with each sample. The cases classified with the greatest uncertainty lie far from the corners of the triangles. These points are brushed **(top left)**, and we examine their location using the linked tour plot **(top right)**. The introduction of linoarach **(bottom)** eliminates the confusion between Sardinia and the North.

Each tree used a random sample of four of the eight variables, as well as a random sample of about a third of the 572 cases. The votes are displayed in the left-hand plot of Fig. 4.11, next to a projection from a 2D tour. Since there are three classes, the votes form a triangle, with one vertex for each region, with oils from the South at the far right, Sardinian oils at the top, and Northern oils at the lower left. Samples that are consistently classified correctly are close to the vertices; cases that are commonly misclassified are further from a vertex. Although forests perfectly classify this data, the number of points falling between the Northern and the Sardinian vertices suggests some potential for error in classifying future samples.

For more understanding of the votes, we turn to another diagnostic: variable importance. Forests return two measures of variable importance, both of which give similar results. Based on the Gini measure, the most important variables, in order, are eicosenoic, linoleic, oleic, palmitic, arachidic, palmitoleic, linolenic, and stearic.

Some of this ordering is as expected, given the initial graphical inspection of the data (Sect. 4.2). The importance of eicosenoic was our first discovery, as shown in the top row of Fig. 4.3. And yes, linoleic is next in importance: The first two plots in Fig. 4.4 make that clear. The surprise is that the forest should consider arachidic to be less important than palmitic. This is not what we found, as shown in the right-hand plot in that figure.

Did we overlook something important in our earlier investigation? We return to the use of the manual manipulation of the tour to see whether palmitic does in fact perform better than arachidic at finding a gap between the two regions. But it does not. By overlooking the importance of arachidic, the random forest never finds an adequate gap between the oils of the Northern and the Sardinian regions, and that probably explains why there is more confusion about some Northern samples than there should be.

We rebuild the forest using a new variable constructed from a linear combination of linoleic and arachidic (linoarach), just as we did when applying the single tree classifier. Since correlated variables reduce each other's importance, we need to remove linoleic and oleic when we add linoarach. Once we have done this, the confusion between Northern and Sardinian oils disappears (Fig. 4.11, lower plot): The points are now tightly clumped at each vertex, which indicates more certainty in their class predictions. The new variable becomes the second most important variable according to the importance diagnostic.

Classifying the oils by the three large regions is too easy a problem for forests; they are designed to tackle more challenging classification tasks. We will use them to examine the oils from the areas in the Southern region (North and South Apulia, Calabria, and Sicily). Remember the initial graphical inspection of the data, which showed that oils from the four areas were not completely separable. The samples from Sicily overlapped those of the three other areas. We will use a forest classifier to see how well it can differentiate the Southern oils by area:

86   4 Supervised Classification

```
> d.olive.sth <- subset(d.olive, region==1,
    select=area:eicosenoic)
> olive.rf <- randomForest(as.factor(area)~.,
    data=d.olive.sth, importance=TRUE, proximity=TRUE,
    mtry=2, ntree=1500)
> order(olive.rf$importance[,5], decreasing=T)
[1] 5 2 4 3 1 6 7 8
> pred <- as.numeric(olive.rf$predicted)
> table(d.olive.sth[,1], olive.rf$predicted)
      1   2   3   4
  1  22   2   0   1
  2   0  53   2   1
  3   0   1 202   3
  4   3   4   5  24
> margin <- olive.rf$vote
> colnames(margin) <- c("Vote1", "Vote2", "Vote3", "Vote4")
> d.olive.rf <- cbind(pred, margin, d.olive.sth)
> gd <- ggobi(d.olive.rf)[1]
> glyph_color(gd) <- c(6,3,2,9)[d.olive.rf$area]
```

After experimenting with several input parameters, we show the results for a forest of 1,500 trees, sampling two variables at each tree node, and yielding an error rate of 0.068. The misclassification table is:

		Predicted area				Error
		North Apulia	Calabria	South Apulia	Sicily	
area	North Apulia	22	2	0	1	0.120
	Calabria	0	53	2	1	0.054
	South Apulia	0	1	202	3	0.019
	Sicily	3	4	5	24	0.333
						0.068

The error of the forest is surprisingly low, but the error is definitely not uniform across classes. Predictions for Sicily are wrong about a third of the time. Figure 4.12 shows some more interesting aspects of the results. For this figure, the following table describes the correspondence between area and symbol:

area	symbol
North Apulia	orange +
Calabria	red +
South Apulia	pink ×
Sicily	yellow ×

Look first at the top row of the figure. The misclassification table is represented by a jittered scatterplot, at the left. A plot from a 2D tour of the four voting variables is in the center. Because there are four groups, the votes lie on a 3D tetrahedron (a simplex). The votes from three of the areas are pretty well separated, one at each "corner," but those from Sicily overlap all of them. Remember that when points are clumped at the vertex, class members are consistently predicted correctly. Since this does not occur for Sicilian oils, we see that there is more uncertainty in the predictions for this area.

The plot at right confirms this observation. It is a projection from a 2D tour of the four most important variables, showing a pattern we have seen before. We can achieve pretty good separation of the oils from North Apulia, Calabria, and South Apulia, but the oils from Sicily overlap all three clusters. Clearly these are tough samples to classify correctly.

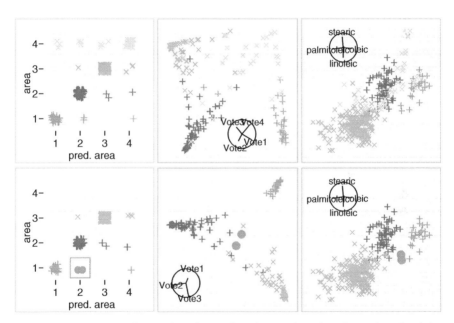

Fig. 4.12. Examining the results of a random forest after classifying the oils of the South by area. A representation of the misclassification table (**left**) is linked to plots of the votes (**middle**) and a 2D tour (**right**). The Sicilian oils have been excluded from the plots in the bottom row.

We remove the Sicilian oils from the plots so we can focus on the other three areas (bottom row of plots). The points representing North Apulian oils form a very tight cluster at a vertex, with three exceptions. Two of these points are misclassified as Calabrian, and we have highlighted them as large filled circles by painting the misclassification plot.

The pattern of the votes (middle plot) suggests that there is high certainty in the predictions for North Apulian oils, with the exception of these two samples. When we watch the votes in the tour for a while, we see that the votes of these two samples travel as if they were in a cluster all their own, which is distinct from the remaining North Apulian oils.

However, when we look at the data, we find the votes for these two samples a bit puzzling. We watch the four most important variables in the tour for a while (as in the right plot), and these two points do not behave as if they were in a distinct cluster; they travel with the rest of the samples from North Apulia. They do seem to be outliers with respect their class, but they are not so far from their group — it is a bit surprising that the forest has trouble classifying these cases.

Rather than exploring the other misclassifications, we leave that for the reader.

In summary, a random forest is a useful method for tackling tough classification problems. Its diagnostics provide a rich basis for graphical exploration, which helps us to digest and evaluate the solution.

4.3.4 Neural networks

Neural networks for classification can be thought of as additive models where explanatory variables are transformed, usually through a logistic function, added to other explanatory variables, transformed again, and added again to yield class predictions. Aside from the data mining literature, mentioned earlier, a good comprehensive and accessible description for statisticians can be found in Cheng & Titterington (1994). The model can be formulated as:

$$\hat{y} = f(x) = \phi(\alpha + \sum_{h=1}^{s} w_h \phi(\alpha_h + \sum_{i=1}^{p} w_{ih} x_i))$$

where x is the vector of explanatory variable values, y is the target value, p is the number of variables, s is the number of nodes in the single hidden layer, and ϕ is a fixed function, usually a linear or logistic function. This model has a single hidden layer and univariate output values. The model is fit by minimizing the sum of squared differences between observed values and fitted values, and the minimization does not always converge. A neural network is a black box that accepts inputs, computes, and spits out predictions. With graphics, some insight into the black box can be gained. We use the feed-forward neural network provided in the nnet package of R (Venables & Ripley 2002) to illustrate.

We continue to work with Olive Oils, and we look at the performance of the neural network in classifying the oils in the four areas of the South, a difficult challenge. Because the software does not include a method for computing the predictive error, we break the data into training and test samples so we can better estimate the predictive error. (We could tweak the neural network to perfectly fit all the data, but then we could not estimate how well it would perform with new data.)

```
> indx.tst <- c(1,7,12,15,16,22,27,32,34,35,36,41,50,54,61,
  68,70,75,76,80,95,101,102,105,106,110,116,118,119,122,134,
  137,140,147,148,150,151,156,165,175,177,182,183,185,186,
  187,190,192,194,201,202,211,213,217,218,219,225,227,241,
  242,246,257,259,263,266,274,280,284,289,291,292,297,305,
  310,313,314,323,330,333,338,341,342,347,351,352,356,358,
  359,369,374,375,376,386,392,405,406,415,416,418,420,421,
  423,426,428,435,440,451,458,460,462,466,468,470,474,476,
  480,481,482,487,492,493,500,501,509,519,522,530,532,541,
  543,545,546,551,559,567,570)
> d.olive.train <- d.olive[-indx.tst,]
> d.olive.test <- d.olive[indx.tst,]
> d.olive.sth.train <- subset(d.olive.train, region==1,
  select=area:eicosenoic)
> d.olive.sth.test <- subset(d.olive.test, region==1,
  select=area:eicosenoic)
```

After trying several values for s, the number of nodes in the hidden layer, we chose $s = 4$; we also chose a linear ϕ, $decay = 0.005$, and $range = 0.06$. We fit the model using many different random starting values, rejecting the results until it eventually converged to a solution with a reasonably low error:

```
> library(nnet)
> olive.nn <- nnet(as.factor(area)~., d.olive.sth.train,
  size=4, linout=T, decay=0.005, range=0.06, maxit=1000)
> targetr <- class.ind(d.olive.sth.train[,1])
> targets <- class.ind(d.olive.sth.test[,1])
> test.cl <- function(true, pred){
      true <- max.col(true)
      cres <- max.col(pred)
      table(true, cres)
  }
> test.cl(targetr, predict(olive.nn,
  d.olive.sth.train[,-1]))
      cres
true   1   2   3   4
   1  16   0   1   2
   2   0  42   0   0
   3   0   1 155   2
   4   1   1   1  24
```

```
> toct.cl(targets, predict(olive.nn, d.olive.sth.test[,-1]))
     cres
true  1  2  3  4
   1  3  2  0  1
   2  0 12  2  0
   3  0  2 45  1
   4  1  2  1  5
> parea <- c(max.col(predict(olive.nn,
    d.olive.sth.train[,-1])),
    max.col(predict(olive.nn, d.olive.sth.test[,-1])))
> d.olive.nn <- cbind(rbind(d.olive.sth.train,
    d.olive.sth.test), parea)
> gd <- ggobi(d.olive.nn)[1]
> glyph_color(gd) <- c(6,3,2,9)[d.olive.nn$area]
```

Below are the misclassification tables for the training and test samples.

Training:

		Predicted area				Error
		North Apulia	Calabria	South Apulia	Sicily	
area	North Apulia	16	0	1	2	0.158
	Calabria	0	42	0	0	0.000
	South Apulia	0	1	155	2	0.019
	Sicily	1	1	1	24	0.111
						0.037

Test:

		Predicted area				Error
		North Apulia	Calabria	South Apulia	Sicily	
area	North Apulia	3	2	0	1	0.333
	Calabria	0	12	2	0	0.143
	South Apulia	0	2	45	1	0.063
	Sicily	1	2	1	5	0.444
						0.156

The training error is $9/246 = 0.037$, and the test error is $12/77 = 0.156$. The overall errors, as in the random forest model, are not uniform across classes. This is particularly obvious in the test error table: The error in classifying North Apulian oils is close to a third, and it is even worse for Sicilian oils, which have an almost even chance of being misclassified.

Our exploration of the misclassifications is shown in Fig. 4.13. (The troublesome Sicilian oils have been excluded from all plots in this figure.) Consider first the plots in the top row. The left-hand plot shows the misclassification table. Two samples of oils from North Apulia (orange +) have been incorrectly classified as South Apulian (pink ×), and these two points have been brushed as filled orange circles. Note where these points fall in the next two plots, which are linked 2D tour projections. One of the two misclassified points is on the edge of the cluster of North Apulian points, close to the Calabrian cluster. It is understandable that there might be some confusion about this case. The other sample is on the outer edge of the North Apulian cluster, but it is far from the Calabrian cluster — this should not have been confused.

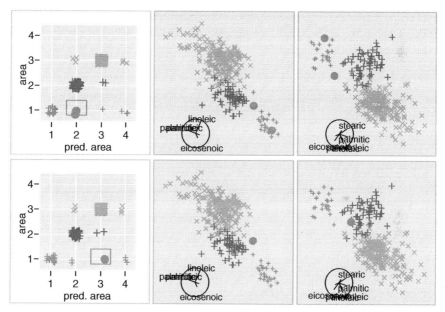

Fig. 4.13. Misclassifications of a feed-forward neural network classifying the oils from the South by area. A representation of the misclassification table (**left column**) is linked to projections viewed in a 2D tour. Different misclassifications are examined in the top and bottom rows. (The Sicilian oils, which would have appeared in the top row of the misclassification tables, have been removed from all plots.)

In the bottom row of plots, we follow the same procedure to examine the single North Apulian sample misclassified as South Apulian. It is painted as a filled orange circle in the misclassification plot and viewed in a tour. This point is on the outer edge of the North Apulian cluster, but it is closer to the Calabrian cluster than the South Apulian cluster. It would be understandable for it to be misclassified as Calabrian, so it is puzzling that it is misclassified as South Apulian.

92   4 Supervised Classification

In summary, a neural network is a black box method for tackling tough classification problems. It will generate different solutions each time the net is fit, some much better than others. When numerical measures suggest that a reasonable model has been found, graphics can be used to inspect the model in more detail.

4.3.5 Support vector machine

A support vector machine (SVM) (Vapnik 1999) is a binary classification method. An SVM looks for gaps between clusters in the data, based on the extreme observations in each class. In this sense it mirrors the graphical approach described at the start of this chapter, in which we searched for gaps between groups. We describe this method more fully than we did the other algorithms for two reasons: first, because of its apparent similarity to the graphical approach, and second, because it is difficult to find a simple explanation of the method in the literature.

The algorithm takes an $n \times p$ data matrix, where each column is scaled to $[-1,1]$ and each row is labeled as one of two classes ($y_i = +1$ or -1), and finds a hyperplane that separates the two groups, if they are separable. Each row of the data matrix is a vector in p-dimensional space, denoted as

$$\mathbf{X} = \begin{bmatrix} x_1 \\ x_2 \\ \vdots \\ x_p \end{bmatrix}$$

and the separating hyperplane can be written as

$$\mathbf{W}'\mathbf{X} + b = 0$$

where $\mathbf{W} = [w_1 \ w_2 \ \ldots \ w_p]'$ is the normal vector to the separating hyperplane and b is a constant. The best separating hyperplane is found by maximizing the margin of separation between the two classes as defined by two parallel hyperplanes:

$$\mathbf{W}'\mathbf{X} + b = 1, \quad \mathbf{W}'\mathbf{X} + b = -1.$$

These hyperplanes should maximize the distance from the separating hyperplane and have no points between them, capitalizing on any gap between the two classes. The distance from the origin to the separating hyperplane is $|b|/||\mathbf{W}||$, so the distance between the two parallel margin hyperplanes is $2/||\mathbf{W}|| = 2/\sqrt{w_1^2 + \ldots + w_p^2}$. Maximizing this is the same as minimizing $||\mathbf{W}||/2$. To ensure that the two classes are separated, and that no points lie between the margin hyperplanes we need:

$$\mathbf{W}'\mathbf{X}_i + b \geq 1, \quad \text{or} \quad \mathbf{W}'\mathbf{X}_i + b \leq -1 \quad \forall i = 1, ..., n$$

which corresponds to

$$y_i(\mathbf{W}'\mathbf{X}_i + b) \geq 1 \quad \forall i = 1, ..., n \tag{4.2}$$

Thus the problem corresponds to

minimizing $\frac{||\mathbf{W}||}{2}$ *subject to* $y_i(\mathbf{X}_i\mathbf{W} + b) \geq 1 \quad \forall i = 1, ..., n.$

Interestingly, only the points closest to the margin hyperplanes are needed to define the separating hyperplane. We might think of these points as lying on or close to the convex hull of each cluster in the area where the clusters are nearest to each other. These points are called support vectors, and the coefficients of the separating hyperplane are computed from a linear combination of the support vectors $\mathbf{W} = \sum_{i=1}^{s} y_i \alpha_i \mathbf{X}_i$, where s is the number of support vectors. We could also use $\mathbf{W} = \sum_{i=1}^{n} y_i \alpha_i \mathbf{X}_i$, where $\alpha_i = 0$ if $\mathbf{X}_i$ is not a support vector. For a good fit the number of support vectors s should be small relative to n. Fitting algorithms can achieve gains in efficiency by using only samples of the cases to find suitable support vector candidates; this approach is used in the SVMLight (Joachims 1999) software.

In practice, the assumption that the classes are completely separable is unrealistic. Classification problems rarely present a gap between the classes, such that there are no misclassifications. Cortes & Vapnik (1995) relaxed the separability condition to allow some misclassified training points by adding a tolerance value ϵ_i to Equation 4.2, which results in the modified criterion $y_i(\mathbf{W}'\mathbf{X}_i + b) > 1 - \epsilon_i, \epsilon_i \geq 0$. Points that meet this criterion but not the stricter one are called slack vectors.

Nonlinear classifiers can be obtained by using nonlinear transformations of $\mathbf{X}_i$, $\phi(\mathbf{X}_i)$ (Boser, Guyon & Vapnik 1992), which is implicitly computed during the optimization using a kernel function K. Common choices of kernels are linear $K(\mathbf{x}_i, \mathbf{x}_j) = \mathbf{x}_i'\mathbf{x}_j$, polynomial $K(\mathbf{x}_i, \mathbf{x}_j) = (\gamma\mathbf{x}_i'\mathbf{x}_j + r)^d$, radial basis $K(\mathbf{x}_i, \mathbf{x}_j) = \exp(-\gamma||\mathbf{x}_i - \mathbf{x}_j||^2)$, or sigmoid functions $K(\mathbf{x}_i, \mathbf{x}_j) = \tanh(\gamma\mathbf{x}_i'\mathbf{x}_j + r)$, where $\gamma > 0, r,$ and d are kernel parameters.

The ensuing minimization problem is formulated as

$$\textit{minimizing } \frac{1}{2}||\mathbf{W}|| + C\sum_{i=1}^{n} \epsilon_i \textit{ subject to } y_i(\mathbf{W}'\phi(\mathbf{X}) + b) > 1 - \epsilon_i$$

where $\epsilon_i \geq 0$, $C > 0$ is a penalty parameter guarding against over-fitting the training data and ϵ controls the tolerance for misclassification. The normal to the separating hyperplane $\mathbf{W}$ can be written as $\sum_{i=1}^{n} y_i\alpha_i\phi(\mathbf{X}_i)$, where points other than the support and slack vectors will have $\alpha_i = 0$. Thus the optimization problem becomes

94   4 Supervised Classification

$$minimizing\ \frac{1}{2}\sum_{i=1}^{n}\sum_{j=1}^{n}y_iy_j\alpha_i\alpha_j K(\mathbf{X}_i,\mathbf{X}_j)+C\sum_{i=1}^{n}\epsilon_i$$

$$subject\ to\ y_i(\mathbf{W}'\phi(\mathbf{X})+b)>1-\epsilon_i$$

We use the svm function in the e1071 package (Dimitriadou, Hornik, Leisch, Meyer & Weingessel 2006) of R, which uses libsvm (Chang & Lin 2006), to classify the oils of the four areas in the Southern region. SVM is a binary classifier, but this algorithm overcomes that limitation by comparing classes in pairs, fitting six separate classifiers, and then using a voting scheme to make predictions. To fit the SVM we also need to specify a kernel, or rely on the internal tuning tools of the algorithm to choose this for us. Automatic tuning in the algorithm chooses a radial basis, but we found that a linear kernel performed better, so that is what we used. (This accords with our earlier visual inspection of the data in Sect. 4.2.) Here is the R code used to fit the model:

```
> library(e1071)
> olive.svm <- best.svm(factor(area) ~ ., data=d.olive.train)
> olive.svm <- svm(factor(area) ~ ., data=d.olive.sth.train,
    type="C-classification", kernel="linear")
> table(d.olive.sth.train[,1], predict(olive.svm,
    d.olive.sth.train))

      1   2   3   4
  1  19   0   0   0
  2   0  42   0   0
  3   0   0 155   3
  4   1   2   3  21
> table(d.olive.sth.test[,1], predict(olive.svm,
    d.olive.sth.test))

      1   2   3   4
  1   6   0   0   0
  2   1  12   1   0
  3   0   0  46   2
  4   1   1   0   7
> support.vectors <- olive.svm$index[
    abs(olive.svm$coefs[,1])<1 &
    abs(olive.svm$coefs[,2])<1 & abs(olive.svm$coefs[,3])<1]
> pointtype <- rep(0,323) # training
> pointtype[247:323] <- 1 # test
> pointtype[olive.svm$index] <- 2 # slack vectors
> pointtype[support.vectors] <- 3 # support vectors
> parea <- c(predict(olive.svm, d.olive.sth.train),
    predict(olive.svm, d.olive.sth.test))
```

```
> d.olive.svm <- cbind(rbind(d.olive.sth.train,
    d.olive.sth.test), parea, pointtype)
> gd <- ggobi(d.olive.svm)[1]
> glyph_color(gd) <- c(6,3,2,9)[d.olive.svm$area]
```

These are our misclassification tables:

Training:

		Predicted area				Error
		North Apulia	Calabria	South Apulia	Sicily	
area	North Apulia	19	0	0	0	0.000
	Calabria	0	42	0	0	0.000
	South Apulia	0	0	155	3	0.019
	Sicily	1	2	3	21	0.222
						0.037

Test:

		Predicted area				Error
		North Apulia	Calabria	South Apulia	Sicily	
area	North Apulia	6	0	0	0	0.000
	Calabria	1	12	1	0	0.143
	South Apulia	0	0	46	2	0.042
	Sicily	1	1	0	7	0.286
						0.078

The training error is $9/246 = 0.037$, and the test error is $6/77 = 0.078$. (The training error is the same as that of the neural network classifier, but the test error is lower.) Most error is associated with Sicily, which we have seen repeatedly to be an especially difficult class to separate. In the training data there are no other errors, and in the test data there are just two samples from Calabria mistakenly classified. Figure 4.14 illustrates our examination of the misclassified cases, one in each row of the figure. (Points corresponding to Sicily were removed from all four plots.) Each of the two cases is brushed (using a filled red circle) in the plot of misclassification table and viewed in a linked 2D tour. Both of these cases are on the edge of their clusters so the confusion of classes is reasonable.

The linear SVM classifier uses 20 support vectors and 29 slack vectors to define the separating planes between the four areas. It is interesting to examine which points are selected as support vectors, and where they are located in the data space. For each pair of classes, we expect to find some

96   4 Supervised Classification

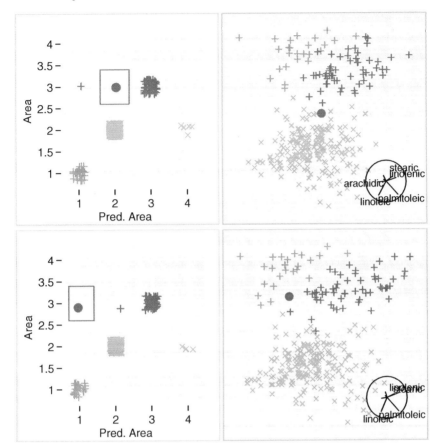

Fig. 4.14. Misclassifications of a support vector machine classifying the oils of the South by area. The misclassification table **(left)** is linked to 2D tour plots **(right)**; different misclassifications are examined in each row of plots. (The oils from Sicily, the fourth area, have been removed from all plots.)

Fig. 4.15. Using the tour to examine the choice of support vectors when classifying Southern oils by area. Support vectors are open circles, and slack vectors are open rectangles; the data points are represented by +es and ×es.

projection in which the support vectors line up on either side of the margin of separation, whereas the slack vectors lie closer to the boundary, perhaps mixed in with the points of other classes.

The plots in Fig. 4.15 represent our use of the 2D tour, augmented by manual manipulation, to look for these projections. (The Sicilian points are again removed.) The support vectors are represented by open circles and the slack vectors by open rectangles, and we have been able to find a number of projections in which the support vectors are on the opposing outer edge of the point clouds for each cluster.

The linear SVM does a very nice job with this difficult classification. The accuracy is almost perfect on three classes, and the misclassifications are quite reasonable mistakes, being points that are on the extreme edges of their clusters. However, this method joins the list of those defeated by the difficult problem of distinguishing the Sicilian oils from the rest.

4.3.6 Examining boundaries

For some classification problems, it is possible to get a good picture of the boundary between two classes. With LDA and SVM classifiers the boundary is described by the equation of a hyperplane. For others the boundary can be determined by evaluating the classifier on points sampled in the data space, using either a regular grid or some more efficient sampling scheme.

We use the R package classifly (Wickham 2006a) to generate points illustrating boundaries, add those points to the original data, and display them in GGobi. Figure 4.16 shows projections of boundaries between pairs of classes in the Olive Oils. In each example, we used the 2D tour with manual control to focus the view on a projection that revealed the boundary between two groups.

```
> library(classifly)
> d.olive.sub <- subset(d.olive,region!=1,
    select=c(region,palmitic:eicosenoic))
> classifly(d.olive.sub, region~linoleic+oleic+arachidic,
    lda)
> classifly(d.olive.sub, region~linoleic+oleic+arachidic,
    svm, probability=TRUE, kernel="linear")
```

The top two plots show tour projections of the North (purple) and Sardinia (green) oils where the two classes are separated and the boundary appears in gray. The LDA boundary (shown at left) slices too close to the Northern oils. This might be due to the violation of the LDA assumption that the two groups have equal variance; since that is not true here, it places the boundary too close to the group with the larger variance. The SVM boundary (at right) is a bit closer to the Sardinian oils than the LDA boundary is, yet it is still a tad too close to the oils from the North.

98   4 Supervised Classification

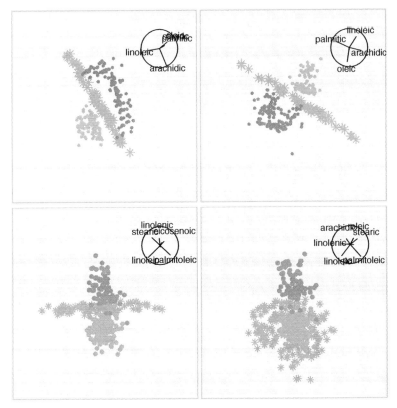

Fig. 4.16. Classification boundaries for different models shown in the 2D tour. Points on the boundary are gray stars. We first compare LDA **(top left)** with linear SVM **(top right)** in finding the boundary between oils from the North and Sardinia. Both boundaries are too close to the cluster of Northern oils. We also compare linear SVM **(bottom left)** and radial kernel SVM **(bottom right)** in finding the boundary between oils from South Apulia and other Southern oils.

The bottom row of plots examines the more difficult classification of the areas of the South, focusing on separating the South Apulian oils (in pink), which is the largest sample, from the oils of the other areas (all in orange). Perfect separation between the classes does not occur. Both plots are tour projections showing SVM boundaries, the left plot generated by a linear kernel and the right one by a radial kernel. Recall that the radial kernel was selected automatically by the SVM software we used, whereas we actually chose to use a linear kernel. These pictures illustrate that the linear basis yields a more reasonable boundary between the two groups. The shape of the clusters of the two groups is approximately the same, and there is only a small overlap of the two. The linear boundary fits this structure neatly. The radial kernel wraps around the South Apulian oils.

4.4 Recap

These partial analyses of the Italian Olive Oils demonstrate that it is possible to get a good mental image of cluster structure in relation to class identity in high-dimensional space. This is possible with many multivariate datasets. Having a good mental image of the class structure can help with many tasks in a classification analysis: choosing an appropriate classifier, validating (or rejecting!) the results of a classification, and simplifying the final model.

The Olive Oils data has nine classes. Jumping straight into classifying the oils into nine classes by area would have led to dismal results. Instead, we aggregated areas to form a new class variable, region, with only three levels. That allowed us to start with the simpler problem of classifying the oils into only three classes, and then use the hierarchical nature of the classes to structure the analysis. This strategy can often be used to simplify classification problems with many classes: divide and conquer.

Graphics can be used to check whether the variance–covariance structure is consistent with a multivariate normal model for classical classifiers, or whether the separations between groups fall along single variable axes so that trees can be used effectively. Linked plots allow us to examine the rich diagnostics provided by random forests and to explore misclassifications exposed by the misclassification table. We can see how well the support vectors mark the separation between classes. It can be surprising to examine the boundary generated by a classifier, even when it has an extremely low error rate.

For the Olive Oils, we saw that the data has a story to tell: The olive oils of Italy are remarkably different in composition based on geographic boundaries. There is something fishy about the Sicilian oils in this data, and the most plausible story is that the Sicilian oils used borrowed olives from neighboring areas. This is interesting! Data analysis is detective work.

Exercises

1. For the Flea Beetles:
   a) Generate a scatterplot matrix. Which variables would contribute to separating the three species?
   b) Generate a parallel coordinate plot. Characterize the three species by the pattern of their traces.
   c) Watch the data in a grand tour. Stop the tour when you see a separation and describe the variables that contribute to the separation.
   d) Using the projection pursuit guided tour with the holes index, find a projection that neatly separates all three species. Put the axes onto the plot, and explain the variables that are contributing to the separation. Using univariate plots, confirm that these variables are important to separate species. (Hint: Transform the data into principal components and enter these variables into the projection pursuit guided tour running the holes index.)

2. For the Australian crabs:
   a) From univariate plots assess whether any individual variables are good classifiers of crabs by species or sex.
   b) From either a scatterplot matrix or pairwise plots, determine which pairs of variables best distinguish the crabs by Species and by sex within species.
   c) Examine the parallel coordinate plot of the five measured variables. Why is a parallel coordinate plot not helpful in determining the importance of variables for this data?
   d) Using Tour1D (and perhaps projection pursuit with the LDA index), find a 1D projection that mostly separates the crabs by species. Report the projection coefficients.
   e) Now transform the five measured variables into principal components and run Tour1D on these new variables. Can you find a better separation of the crabs by species?
3. For the Italian Olive Oils:
   a) Split the samples from North Italy into 2/3 training and 1/3 test samples for each area.
   b) Build a tree model to classify the oils by area for the three areas of North Italy. Which are the most important variables? Make plots of these variables. What is the accuracy of the model for the training and test sets?
   c) Build a random forest to classify oils into the three areas of North Italy. Compare the order of importance of variables with what you found from a single tree. Make a parallel coordinate plot in the order of variable importance.
   d) Fit a support vector machine model and a feed-forward neural network model to classify oils by area for the three areas of the North. Using plots, compare the predictions of each point for SVM, a feed-forward neural network and random forests.
4. For the TAO data:
   a) Build a classifier to distinguish between the normal and the El Niño years. Depending on the classifier you use, you may need to impute the missing values first.
   b) Which variables are important for distinguishing an El Niño year from a normal year?
5. For spam:
   a) Create a new variable domain.reduced that reduces the number of categories of domain to "edu," "com," "gov," "org," "net," and "other."
   b) Using spam as the class variable, and using explanatory variables day of week, time of day, size.kb, box, domain.reduced, local, digits, name, capct, special, credit, sucker, porn, chain, username, and large text, build a random forest classifier using $mtry = 2$.
   c) What is the order of importance of the variables?
   d) How many non-spam emails are misclassified as spam?

e) Examine a scatterplot of predicted class against actual class, using jittering to spread the values, and a parallel coordinate plot of the explanatory variables in the order of importance returned by the forest. Brush the cases corresponding to non-spam email that has been predicted to be spam. Characterize these email messages (e.g., all from the local box, small number of digits). Now look at the emails that are spam and correctly classified as spam. Is there something special about them?

f) Examine the relationship between Spam (actual class) and Spam.Prob (probability of being spam as estimated by Iowa State University's mail administrators). How many cases that are not spam are rated as more than 50% likely to be spam?

g) Examine the probability rating for cases corresponding to non-spam that your random forest classified as spam. Write a description of the email that has the highest probability of being spam and is also considered to be very likely to be spam by random forests.

h) Which user has the highest proportion of non-spam email classified as spam?

i) Based on your exploration of this data, which variables would you suggest are the most important in determining if an email message is spam?

6. In this exercise, your goal is to build a classifier for the music data that will distinguish between rock and classical tracks.

   a) This data has 70 explanatory variables for 62 samples. Reduce the number of variables to fewer than 10, choosing those that are the most suitable candidates on which to build a classifier. (Hint: One of the problems to consider is that there are several missing values. It might be possible to reduce the number of variables in a way that also fixes the missing values problem.)

   b) Split the data into two samples, with 2/3 of the data in the training sample and 1/3 in the test sample. Report which cases are in each sample.

   c) Build your best classifier for distinguishing rock from classical tracks.

   d) Predict the five new tracks as either rock or classical.

5
Cluster Analysis

The aim of unsupervised classification, or cluster analysis, is to organize observations into similar groups. Cluster analysis is a commonly used, appealing, and conceptually intuitive, statistical method. Some of its uses include market segmentation, where customers are grouped into clusters with similar attributes for targeted marketing; gene expression analysis, where genes with similar expression patterns are grouped together; and the creation of taxonomies of animals, insects, or plants. A cluster analysis results in a simplification of a dataset for two reasons: first, because the dataset can be summarized by a description of each cluster, and second, because each cluster, which is now relatively homogeneous, can be analyzed separately. Thus, it can be used to effectively reduce the size of massive amounts of data.

Organizing objects into groups is a task that seems to come naturally to humans, even to small children, and perhaps this is why it is an apparently intuitive method in data analysis. However, cluster analysis is more complex than it initially appears. Many people imagine that it will produce neatly separated clusters like those in the top left plot of Fig. 5.1, but it almost never does. Such ideal clusters are rarely encountered in real data, so we often need to modify our objective from "find the natural clusters in this data" to "organize the cases into groups that are similar in some way." Even though this may seem disappointing when compared with the ideal, it is still often an effective means of simplifying and understanding a dataset.

At the heart of the clustering process is the work of discovering which variables are most important for defining the groups. It is often true that we only require a subset of the variables for finding clusters, whereas another subset (called *nuisance variables*) has no impact. In the bottom left plot of Fig. 5.1, it is clear that the variable plotted horizontally is important for splitting this data into two clusters, whereas the variable plotted vertically is a nuisance variable. Nuisance is an apt term for these variables, because they can radically change the interpoint distances and impair the clustering process.

104   5 Cluster Analysis

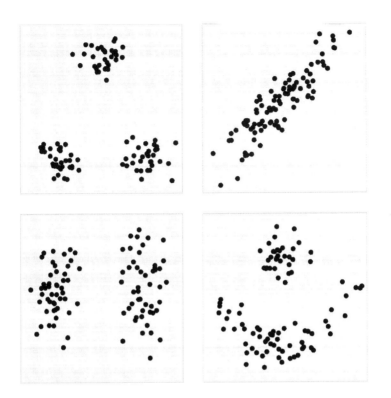

Fig. 5.1. Different structures in data and their impact on cluster analysis. When there are well-separated groups **(top left)**, it is simple to group similar observations. Even when there are not **(top right)**, grouping observations may still be useful. There may be nuisance variables that do not contribute to the clustering **(bottom left)**, and there may be oddly shaped clusters **(bottom right)**.

Dynamic graphical methods help us to find and understand the cluster structure in high dimensions. With the tools in our toolbox, primarily tours, along with linked scatterplots and parallel coordinate plots, we can see clusters in high-dimensional spaces. We can detect gaps between clusters, the shape and relative positions of clusters, and the presence of nuisance variables. We can even find unusually shaped clusters, like those in the bottom right plot in Fig. 5.1. In simple situations we can use graphics alone to group observations into clusters, using a "spin and brush" method. In more difficult data problems, we can assess and refine numerical solutions using graphics.

This chapter discusses the use of interactive and dynamic graphics in the clustering of data. Section 5.1 introduces cluster analysis, focusing on interpoint distance measures. Section 5.2 describes an example of a purely graphical approach to cluster analysis, the spin and brush method. In the example shown in that section, we were able to find simplifications of the data that had not been found using numerical clustering methods, and to find a variety of structures in high-dimensional space. Section 5.3 describes methods for reducing the interpoint distance matrix to an intercluster distance matrix using hierarchical algorithms and model-based clustering, and shows how graphical tools are used to assess the results of numerical methods. Section 5.5 summarizes the chapter and revisits the data analysis strategies used in the examples. A good companion to the material presented in this chapter is Venables & Ripley (2002), which provides data and code for practical examples of cluster analysis using R. Section 5.5 summarizes the chapter and revisits the data analysis strategies used in the examples.

5.1 Background

Before we can begin finding groups of cases that are similar, we need to decide on a definition of similarity. How is similarity defined? Consider a dataset with three cases and four variables, described in matrix format as

$$\mathbf{X} = \begin{bmatrix} \mathbf{X}_1 \\ \mathbf{X}_2 \\ \mathbf{X}_3 \end{bmatrix} = \begin{bmatrix} 7.3 & 7.6 & 7.7 & 8.0 \\ 7.4 & 7.2 & 7.3 & 7.2 \\ 4.1 & 4.6 & 4.6 & 4.8 \end{bmatrix}$$

which is plotted in Fig. 5.2. The Euclidean distance between two cases (rows of the matrix) is defined as

$$d_{\text{Euc}}(\mathbf{X}_i, \mathbf{X}_j) = ||\mathbf{X}_i - \mathbf{X}_j|| \qquad i, j = 1, \ldots, n,$$

where $||\mathbf{X}_i|| = \sqrt{X_{i1}^2 + X_{i2}^2 + \ldots + X_{ip}^2}$. For example, the Euclidean distance between cases 1 and 2 in the above data, is

$$\sqrt{(7.3 - 7.4)^2 + (7.6 - 7.2)^2 + (7.7 - 7.3)^2 + (8.0 - 7.2)^2} = 1.0.$$

For the three cases, the interpoint Euclidean distance matrix is

$$d_{\text{Euc}} = \begin{bmatrix} 0.0 & & \\ 1.0 & 0.0 & \\ 6.3 & 5.5 & 0.0 \end{bmatrix} \begin{matrix} \mathbf{X}_1 \\ \mathbf{X}_2 \\ \mathbf{X}_3 \end{matrix}$$

Cases 1 and 2 are more similar to each other than they are to case 3, because the Euclidean distance between cases 1 and 2 is much smaller than the distance between cases 1 and 3 and between cases 2 and 3.

106   5 Cluster Analysis

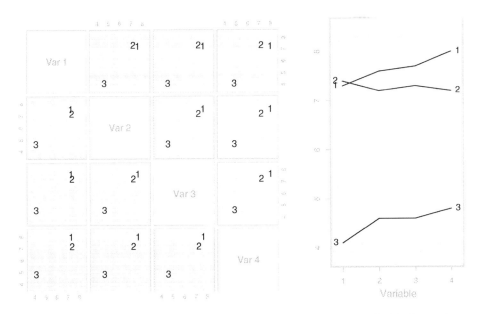

Fig. 5.2. Clustering the example data. The scatterplot matrix **(left)** shows that cases 1 and 2 have similar values. The parallel coordinate plot **(right)** allows a comparison of other structure, which shows the similarity in the profiles on cases 1 and 3.

There are many different ways to calculate similarity. In recent years similarity measures based on correlation distance have become common. Correlation distance is typically used where similarity of structure is more important than similarity in magnitude.

As an example, see the parallel coordinate plot of the sample data at the right of Fig. 5.2. Cases 1 and 3 are widely separated, but their shapes are similar (low, medium, medium, high). Case 2, although overlapping with Case 1, has a very different shape (high, medium, medium, low). The correlation between two cases is defined as

$$\rho(\mathbf{X}_i, \mathbf{X}_j) = \frac{(\mathbf{X}_i - c_i)'(\mathbf{X}_j - c_j)}{\sqrt{(\mathbf{X}_i - c_i)'(\mathbf{X}_i - c_i)}\sqrt{(\mathbf{X}_j - c_j)'(\mathbf{X}_j - c_j)}} \quad (5.1)$$

When c_i, c_j are the sample means $\bar{\mathbf{X}}_i, \bar{\mathbf{X}}_j$, then ρ is the Pearson correlation coefficient. If, indeed, they are set at 0, as is commonly done, ρ is a generalized correlation that describes the angle between the two data vectors. The correlation is then converted to a distance metric; one equation for doing so is as follows:

$$d_{\text{Cor}}(\mathbf{X}_i, \mathbf{X}_j) = \sqrt{2(1 - \rho(\mathbf{X}_i, \mathbf{X}_j))}$$

The above distance metric will treat cases that are strongly negatively correlated as the most distant.

The interpoint distance matrix for the sample data using d_{Cor} and the Pearson correlation coefficient is

$$d_{\text{Cor}} = \begin{bmatrix} 0.0 & & \\ 3.6 & 0.0 & \\ 0.1 & 3.8 & 0.0 \end{bmatrix}$$

By this metric, cases 1 and 3 are the most similar, because the correlation distance is smaller between these two cases than the other pairs of cases.

Note that these interpoint distances differ dramatically from those for Euclidean distance. As a consequence, the way the cases would be clustered is also be very different. Choosing the appropriate distance measure is an important part of a cluster analysis.

After a distance metric has been chosen and a cluster analysis has been performed, the analyst must evaluate the results, and this is actually a difficult task. A cluster analysis does not generate p-values or other numerical criteria, and the process tends to produce hypotheses rather than testing them. Even the most determined attempts to produce the "best" results using modeling and validation techniques may result in clusters that, although seemingly significant, are useless for practical purposes. As a result, cluster analysis is best thought of as an exploratory technique, and it can be quite useful despite the lack of formal validation because of its power in data simplification.

The context in which the data arises is the key to assessing the results. If the clusters can be characterized in a sensible manner, and they increase our knowledge of the data, then we are on the right track. To use an even more pragmatic criterion, if a company can gain an economic advantage by using a particular clustering method to carve up their customer database, then that is the method they should use.

5.2 Purely graphics

A purely graphical spin and brush approach to cluster analysis works well when there are good separations between groups, even when there are marked differences in variance structures between groups or when groups have nonlinear boundaries. It does not work very well when there are clusters that overlap, or when there are no distinct clusters but rather we simply wish to partition the data. In these situations it may be better to begin with a numerical solution and to use visual tools to evaluate it, perhaps making refinements subsequently. Several examples of the spin and brush approach are documented in the literature, such as Cook et al. (1995) and Wilhelm, Wegman & Symanzik (1999).

This description of the spin and brush approach on PRIM7, a particle physics dataset, follows that in Cook et al. (1995). The data contains seven variables. We have no labels for the data, so when we begin, all the points have the same color and glyph. Watch the data in a tour for a few minutes and you will see that there are no natural clusters, but there is clearly structure.

We will use the projection pursuit guided tour to help us find that structure. We will tour on the principal components, rather than the raw variables, because that improves the performance of the projection pursuit indexes. Two indexes are useful for detecting clusters: holes and central mass. The holes index is sensitive to projections where there are few points (i.e., a hole) in the center. The central mass index is the opposite: It is sensitive to projections that have too many points in the center. These indexes are explained in Chap. 2.

The holes index is usually the most useful for clustering, but not for the particle physics data, because it does not have a "hole" at the center. The central mass index is the most appropriate here. Alternate between optimization (a guided tour) and the unguided grand tour to find local maxima, each of which is a projection that is potentially useful for revealing clusters. The process is illustrated in Fig. 5.3.

The top left plot shows the initial default projection, the second principal component plotted against the first. The plot next to it shows the projected data corresponding to the first local maximum found by the guided tour. It has three strands of points stretching out from the central clump and several outliers. We brush the points along each strand, in red, blue, and orange, and we paint the outliers with open circles. (See the next two plots.) We continue by choosing a new random start for the guided tour, and then waiting until new territory in the data is discovered.

The optimization settles on a projection where there are three strands visible, as observed in the leftmost plot in the second row. Two strands have been previously brushed, but a new one has appeared; this is painted yellow.

We also notice that there is another new strand hidden below the red strand. It is barely distinguishable from the red strand in this projection, but the two strands separate widely in other projections. It is tricky to brush it, because it is not well separated in this projection. We use a trick: Hide the red points, brush the new strand green, and "unhide" the red points again (middle plot in the second row).

Five clusters have been easily identified, and now finding new clusters in this data is increasingly difficult. After several more alternations between the grand tour and the guided tour, we find something new (shown in the rightmost plot in the second row): One more strand has emerged, and we paint it pink.

5.2 Purely graphics

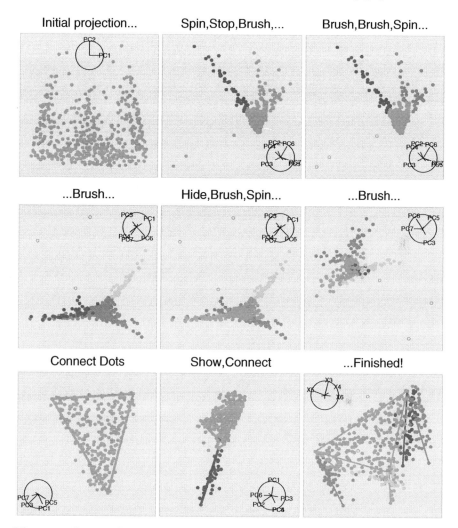

Fig. 5.3. Stages of spin and brush on PRIM7. The high-dimensional geometry emerges as the clusters are painted.

The results at this stage are summarized by the bottom row of plots. There is a very visible triangular component (in gray) revealed when all of the colored points are hidden. We check the shape of this cluster by drawing lines between outer points to contain the inner ones. Touring after the lines are drawn helps to check how well they match the shape of the clusters. The colored groups pair up at each vertex, and we draw in the shape of these too — a single line matches the structures reasonably well.

The final step of the spin and brush clustering is to clean up this solution, touching up the color groups by continuing to tour, and repainting a point here and there When we finish, we have found seven clusters in this data that form a very strong geometric object in the data space: a two-dimensional (2D) triangle, with two one-dimensional (1D) strands extending in different directions from each vertex. The lines confirm our understanding of this object's shape, because the points stay close to the lines in all of the projections observed in a tour.

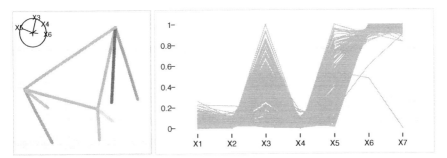

Fig. 5.4. The PRIM7 model summarized. The model summary (**left**) was formed by adding line segments manually. In the parallel coordinate plot, the profiles highlighted in dark gray correspond to the points in the 2D triangle at the center of the model.

The next stage of cluster analysis is to characterize the nature of the clusters. To do that, we would calculate summary statistics for each cluster, and plot the clusters (Fig. 5.4). When we plot the clusters of the particle physics data, we find that the 2D triangle exists primarily in the plane defined by X3 and X5. If you do the same, notice that the variance in measurements for the gray group is large in variables X3 and X5, but negligible in the other variables. The linear pieces can also be characterized by their distributions on each of the variables. With this example, we have shown that it is possible to uncover very unusual clusters in data without any domain knowledge.

Here are several tips about the spin and brush approach.

- Save the dataset frequently during the exploration of a complex dataset, being sure to save your colors and glyphs, because it may take several sessions to arrive at a final clustering.
- Manual controls are useful for refining the optimal projection because another projection in the neighborhood may be more revealing.
- The holes index is usually the most successful projection pursuit index for finding clusters.
- Principal component coordinates may provide a better starting point than the raw variables.

Finally, the spin and brush method will not work well if there are no clear separations in the data, and the clusters are high-dimensional, unlike the low-dimensional clusters found in this example.

5.3 Numerical methods

5.3.1 Hierarchical algorithms

Hierarchical cluster algorithms sequentially fuse neighboring points to form ever-larger clusters, starting from a full interpoint distance matrix. *Distance between clusters* is described by a "linkage method": For example, single linkage uses the smallest interpoint distance between the members of a pair of clusters, complete linkage uses the maximum interpoint distance, and average linkage uses the average of the interpoint distances. A good discussion on cluster analysis can be found in Johnson & Wichern (2002) or Everitt, Landau & Leese (2001).

Figure 5.5 contains several plots that illustrate the results of the hierarchical clustering of the particle physics data; we used Euclidean interpoint distances and the average linkage method. This is computed by:

```
> library(rggobi)
> d.prim7 <- read.csv("prim7.csv")
> d.prim7.dist <- dist(d.prim7)
> d.prim7.dend <- hclust(d.prim7.dist, method="average")
> plot(d.prim7.dend)
```

The dendrogram at the top shows the result of the clustering process. Several large clusters were fused late in the process, with heights (indicated by the height of the horizontal segment connecting two clusters) well above those of the first joins; we will want to look at these. Two points were fused with the rest at the very last stages, which indicates that they are outliers and have been assigned to singleton clusters.

We cut the dendrogram to produce nine clusters because we would expect to see seven clusters and a few outliers based on our observations from the spin and brush approach, and our choice looks reasonable given the structure of the dendrogram. (In practice, we would usually explore the clusters corresponding to several different cuts of the dendrogram.) We assign each cluster an integer identifier, and in the following plots, you see the results of highlighting one cluster at a time and then running the grand tour to focus on the placement of that cluster within the data. This R code follows this sequence of actions:

```
> gd <- ggobi(d.prim7)[1]
> clust9 <- cutree(d.prim7.dend, k=9)
> glyph_color(gd)[clust9==1] <- 9 # highlight triangle
> glyph_color(gd)[clust9==1] <- 1 # reset color
> glyph_color(gd)[clust9==2] <- 9 # highlight cluster 2
```

112   5 Cluster Analysis

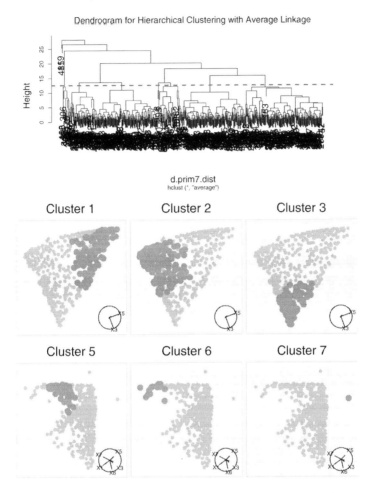

Fig. 5.5. Hierarchical clustering of the particle physics data. The dendrogram shows the results of clustering the data using average linkage. Clusters 1, 2, and 3 carve up the base triangle of the data; clusters 5 and 6 divide one of the arms; and cluster 7 is a singleton.

The top three plots show, respectively, clusters 1, 2, and 3: These clusters roughly divide the main triangular section of the data into three. The bottom row of plots show clusters labeled 5, and 6, which lie along the linear pieces, and cluster 7, which is a singleton cluster corresponding to an outlier in the data.

The results are reasonably easy to interpret. Recall that the basic geometry underlying this data is that there is a 2D triangle with two linear strands extending from each vertex. The hierarchical average linkage clustering of the particle physics data using nine clusters essentially divides the data into three chunks in the neighborhood of each vertex (clusters 1, 2, and 3), three pieces at the ends of the six linear strands (4, 5, and 6), and three clusters containing outliers (7, 8, and 9). This data provides a big challenge for any cluster algorithm — low-dimensional pieces embedded in high-dimensional space — and we are not surprised that no algorithm that we have tried will extract the structure we found using interactive tools.

The particle physics dataset is ill-suited to hierarchical clustering, but this extreme failure is an example of a common problem. When performing cluster analysis, we want to group the observations into clusters without knowing the distribution of the data. How many clusters are appropriate? What do the clusters look like? Could we just as confidently divide the data in several different ways and get very different but equally valid interpretations? Graphics can help us assess the results of a cluster analysis by helping us explore the distribution of the data and the characteristics of the clusters.

5.3.2 Model-based clustering

Model-based clustering (Fraley & Raftery 2002) fits a multivariate normal mixture model to the data. It uses the EM algorithm to fit the parameters for the mean, variance–covariance of each population, and the mixing proportion. The variance–covariance matrix is re-parametrized using an eigen-decomposition

$$\Sigma_k = \lambda_k D_k A_k D'_k, \quad k = 1, \ldots, g \text{ (number of clusters)}$$

resulting in several model choices, ranging from simple to complex:

Name	Σ_k	Distribution	Volume	Shape	Orientation
EII	λI	Spherical	equal	equal	NA
VII	$\lambda_k I$	Spherical	variable	equal	NA
EEI	λA	Diagonal	equal	equal	coordinate axes
VEI	$\lambda_k A$	Diagonal	variable	equal	coordinate axes
VVI	$\lambda_k A_k$	Diagonal	variable	variable	coordinate axes
EEE	$\lambda DAD'$	Ellipsoidal	equal	equal	equal
EEV	$\lambda D_k A D'_k$	Ellipsoidal	equal	equal	variable
VEV	$\lambda_k D_k A D'_k$	Ellipsoidal	variable	equal	variable
VVV	$\lambda_k D_k A_k D'_k$	Ellipsoidal	variable	variable	variable

Note the distribution descriptions "spherical" and "ellipsoidal." These are descriptions of the shape of the variance–covariance for a multivariate normal distribution. A standard multivariate normal distribution has a variance–covariance matrix with zeros in the off-diagonal elements, which corresponds to spherically shaped data. When the variances (diagonals) are different or the variables are correlated, then the shape of data from a multivariate normal is ellipsoidal.

The models are typically scored using the Bayes Information Criterion (BIC), which is based on the log likelihood, number of variables, and number of mixture components. They should also be assessed using graphical methods, as we demonstrate using the Australian Crabs data. We start with two of the five real-valued variables (frontal lobe and rear width) and one species (Blue).

```
> library(mclust)
> d.crabs <- read.csv("australian-crabs.csv")
> d.blue.crabs <- subset(d.crabs,
    species=="Blue", select=c(sex,FL:BD))
```

The goal is to determine whether model-based methods can discover clusters that will distinguish between the two sexes.

Figure 5.6 contains the plots we will use to examine the results of model-based clustering on this reduced dataset. The top leftmost plot shows the data, with male and female crabs distinguished by color and glyph. The two sexes correspond to long cigar-shaped objects that overlap a bit, particularly for smaller crabs. The "cigars" are not perfectly regular: The variance of the data is smaller at small values for both sexes, so that our cigars are somewhat wedge-shaped. The orientation of the longest direction of variance differs slightly between groups too: The association has a steeper slope for female crabs than for males, because female crabs have relatively larger rear width than male crabs. With the heterogeneity in variance–covariance, this data does not strictly adhere to the multivariate normal mixture model underlying model-based methods, but we hope that the departure from regularity is not so extreme that it prevents the model from working.

The top right plot shows the BIC results for a full range of models, EEE, EEV, and VVV variance–covariance parametrization for one to nine clusters:

```
> blue.crabBIC <- mclustBIC(subset(d.blue.crabs,
    select=c(FL,RW)), modelNames=c("EEE","EEV","VVV"))
> blue.crabBIC
 BIC:
         EEE       EEV       VVV
1 -810.3289 -810.3289 -810.3289
2 -820.7272 -778.6450 -783.7705
3 -832.7712 -792.5937 -821.8645
4 -824.8927 -835.5631 -835.7799
5 -805.8402 -805.9425 -853.1395
6 -807.8380 -821.1586 -879.3500
7 -827.1099 -860.7258 -878.0679
8 -833.8051 -861.1460 -891.9757
9 -835.6620 -854.6120 -904.6108
> plot(blue.crabBIC)
 EEE EEV VVV
  15  12   0
```

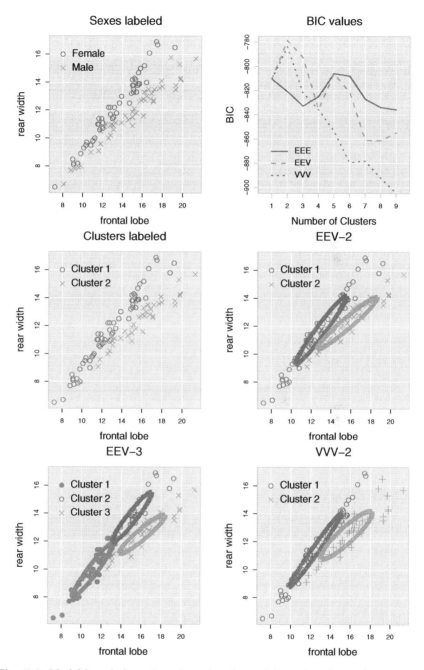

Fig. 5.6. Model-based clustering of a reduced set of Australian Crabs. A scatterplot (**top left**) shows the data with values of sex labeled. A plot of the BIC values for the full range of models (**top right**) shows that the best model organizes the cases into two clusters using EEV parametrization. We label the cases by cluster for the best model (**middle left**). The remaining three plots include ellipses representing the variance–covariance estimates of the three best models, EEV-2 (**middle right**), EEV-3 (**bottom left**), and VVV-2 (**bottom right**).

116   5 Cluster Analysis

The best model, EEV-2, used the equal volume, equal shape, and different orientation variance–covariance parametrization and divided the data into two clusters. This solution seems to be perfect! We can imagine that this result corresponds to two equally shaped ellipses that intersect near the lowest values of the data and angle toward higher values. We will check by drawing ellipses representing the variance–covariance parametrization on the data plots. The parameter estimates are used to scale and center the ellipses:

```
> mclst1 <- mclustBIC(subset(d.blue.crabs,
    select=c(FL,RW)), G=2, modelNames="EEV")
> mclst1
 BIC:
        EEV
2 -778.645
> smry1 <- mclustModel(subset(d.blue.crabs,
    select=c(FL,RW)), mclst1, G=2, modelNames="EEV")
> vc <- smry1$parameters$variance$sigma[,,1]
> xm <- smry1$parameters$mean[,1]
> y1 <- f.vc.ellipse(vc,xm,500)
> ...
```

yielding the plots in the middle and bottom rows of Fig. 5.6. In the plot of the data alone, cluster id is used for the color and glyph of points. (Compare this plot with the one directly above it, in which the classes are known.) Cluster 1 mostly corresponds to the female crabs, and cluster 2 to the males, except that all the small crabs, both male and female, have been assigned to cluster 1. In the rightmost plot, we have added ellipses representing the estimated variance–covariances. The ellipses are the same shape, as specified by the model, but the ellipse for cluster 2 is shifted toward the large values.

The next two best models, according to the BIC values, are EEV-3 and VVV-2. The plots in the bottom row display representations of the variance–covariances for these models. EEV-3 organizes the crabs into three clusters according to the size, not the sex, of the crabs. The VVV-2 solution is similar to EEV-2.

What solution is the best for this data? If the EEV-3 model had done what we intuitively expected, it would have been ideal: The sexes of smaller crabs are indistinguishable, so they should be afforded their own cluster, whereas larger crabs could be clustered into males and females. However, the cluster that includes the small crabs also includes a fair number of middle-sized female crabs.

Finally, model-based clustering did not discover the true gender clusters. Still, it produced a useful and interpretable clustering of the crabs.

Plots are indispensable for choosing an appropriate cluster model. It is easy to visualize the models when there are only two variables but increasingly difficult as the number of variables grows. Tour methods save us from producing page upon page of plots. They allow us to look at many projections of the data, which enables us to conceptualize the shapes and relationships between clusters in more than two dimensions.

Figure 5.7 displays the graphics for the corresponding high-dimensional investigation using all five variables and four classes (two species, two sexes) of the Australian Crabs. The cluster analysis is much more difficult now. Can model-based clustering uncover these four groups?

In the top row of plots, we display the raw data, before modeling. Each plot is a tour projection of the data, colored according to the four true classes. The blue and purple points are the male and female crabs of the blue species, and the yellow and orange points are the male and female crabs of the orange species. This table will help you keep track:

	Male	Female
Blue Species	blue rectangles	purple circles
Orange Species	yellow circles	orange rectangles

The clusters corresponding to the classes are long thin wedges in five dimensions (5D), with more separation and more variability at larger values, as we saw in the subset just discussed. The rightmost plot shows the "looking down the barrel" view of the wedges. At small values the points corresponding to the sexes are mixed (leftmost plot). The species are reasonably well separated even for small crabs (middle plot). The variance–covariance is wedge-shaped rather than elliptical, but again we hope that modeling based on the normal distribution that has elliptical variance–covariance will be adequate.

In the results from model-based clustering, there is very little difference in BIC value for variance–covariance models EEE, EEV, VEV, and VVV, with a number of clusters from three to eight. The best model is EEV-3, and EEV-4 is second best. We know that three clusters is insufficient to capture the four classes we have in mind, so we examine the four-cluster solution.

```
> mclst4 <- mclustBIC(subset(d.blue.crabs,select=c(FL:BD)),
    G=1:8, modelNames=c("EEE","EEV","VVV"))
> plot(mclst4)
EEE EEV VVV
 15  12   0
> mclst5 <- mclustBIC(subset(d.blue.crabs,select=c(FL:BD)),
    G=4, modelNames="EEV")
> smry5 <- mclustModel(subset(d.blue.crabs,select=c(FL:BD)),
    mclst5, G=4, modelNames="EEV")
```

The bottom row of plots in Fig. 5.7 illustrates the four-cluster model in three different projections, matching the projections in the top row showing the data.

```
> vc <- smry5$parameters$variance$sigma[,,1]
> mn <- smry5$parameters$mean[,1]
```

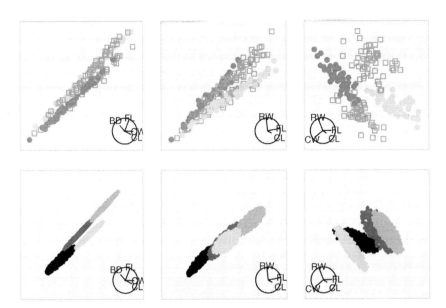

Fig. 5.7. Comparing the Australian Crabs data with results of model-based clustering using all variables. Compare the tour projections of the 5D data **(top row)** with the 5D ellipses corresponding to the variance–covariance in the four-cluster model **(bottom row)**. The ellipses of the four clusters do not match the four known groups in the data.

```
> y1 <- f.vc.ellipse(vc, mn)
> ...
> mclst5.model <- cbind(matrix(NA,500*4,3),
    rbind(y1,y2,y3,y4))
> colnames(mclst5.model) <-
    c("Species","Sex","Index","FL","RW","CL","CW","BD")
> d.crabs.model <- rbind(d.crabs, mclst5.model)
> gd <- ggobi(d.crabs.model)[1]
> glyph_color(gd) <- c(rep(4,50), rep(1,50), rep(9,50),
    rep(6,50), rep(8,2000))
```

In each view, the ellipsoids representing the variance–covariance estimates for the four clusters are shown in four shades of gray, because none of these match any actual cluster in the data. Remember that these are 2D projections of 5D ellipsoids. The resulting clusters from the model do not match the true classes very well. The result roughly captures the two species, as we see in the plots in the first column, where the species are separated both in the data and in the ellipses. On the other hand, the grouping corresponding to sex is completely missed: See the plots in the middle and right-hand columns, where sexes are separated in the actual data but the ellipses are not separated. Just as in the smaller subset (two variables, one species) discussed earlier, there is a cluster for the smaller crabs of both species and sexes. The results of model-based clustering on the full 5D data are very unsatisfactory.

In summary, plots of the data and parameter estimates for model-based cluster analysis are very useful for understanding the solution, and choosing an appropriate model. Tours are very helpful for examining the results in higher dimensions, for arbitrary numbers of variables.

5.3.3 Self-organizing maps

A self-organizing map (SOM) (Kohonen 2001) is constructed using a constrained k-means algorithm. A 1D or 2D net is stretched through the data. The knots, in the net, form the cluster means, and the points closest to the knot are considered to belong to that cluster. The similarity of nodes (and their corresponding clusters) is defined as proportional to their distance from one another on the net.

We will demonstrate SOM using the music data. The data has 62 cases, each one corresponding to a piece of music. For each piece there are seven variables: the artist, the type of music, and five characteristics, based on amplitude and frequency, that were computed using the first 40 seconds of the piece on CD. The music used included popular rock songs by Abba, the Beatles, and the Eels; classical compositions by Vivaldi, Mozart and Beethoven; and several new wave pieces by Enya. Figure 5.8 displays a typical view of the results of clustering using SOM on the music data. Each data point corresponds to a piece of music and is labeled by the band or the composer. The map was generated by this R code:

```
> library(som)
> d.music <- read.csv("music-sub.csv", row.names=1)
> d.music.std <- cbind(subset(d.music.std,
    select=c(artist,type)),
    apply(subset(d.music.std,select=lvar:lfreq),
    2, f.std.data))
> music.som <- som(subset(d.music.std,select=lvar:lfreq),
    6, 6, neigh="bubble", rlen=1000)
```

The left plot in Fig. 5.8 is called the 2D map view. Here we have used a 6×6 net pulled through the 5D data. The net that was wrapped through the high-dimensional space is straightened and laid out flat, and the points, like fish in a fishing net, are laid out where they have been trapped. In the plot shown here, the points have been jittered slightly, away from the knots of the net, so that the labels do not overlap too much. If the fit is good, the points that are close together in this 2D map view are close together in the high-dimensional data space and close to the net as it was placed in the high-dimensional space.

120   5 Cluster Analysis

Much of the structure in the map is no surprise: The rock (purple) and classical tracks (green) are on opposing corners, with rock in the upper right and classical in the lower left. The Abba tracks are all grouped at the top and left of the map. The Beatles and Eels tracks are mixed. There are some unexpected associations: For example, one Beatles song, which turns out to be "Hey Jude," is mixed among the classical compositions!

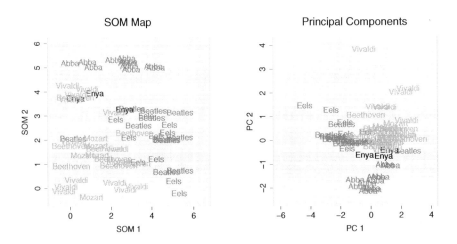

Fig. 5.8. Comparison of clustering music tracks using a self-organizing map versus principal components. The data was clustered using a self-organizing map, as shown in a 2D map view. (When tracks clustered at a node, jittering was used to spread the labels.) Compare with the scatterplot of the first two principal components.

Construction of a self-organizing map is a dimension reduction method, akin to multidimensional scaling (Borg & Groenen 2005) or principal component analysis (Johnson & Wichern 2002). Using principal component analysis to find a low-dimensional approximation of the similarity between music pieces, yields the second plot in Fig. 5.8. There are many differences between the two representations. The SOM has a more even spread of music pieces across the grid, in contrast to the stronger clumping of points in the PCA view. Indeed, the PCA view shows several outliers, notably one of the Vivaldi compositions, which could lead us to learn things about the data that we might miss by relying exclusively on the SOM.

These two methods, SOM and PCA, have provided two contradictory clustering models. How can we determine which is the more accurate description of the data structure? An important part of model assessment is plotting the model in relation to the data. Although plotting the low-dimensional map is the common way to graphically assess the SOM results, it is woefully limited. If a model has flaws, they may not show up in this view and will only appear

5.3 Numerical methods

in plots of the model in the data space. We will use the grand tour to create these plots, and this will help us assess the two models.

We will use a grand tour to view the net wrapped in among the data, hoping to learn how the net converged to this solution, and how it wrapped through the data space. Actually, it is rather tricky to fit a SOM: Like many algorithms, it has a number of parameters and initialization conditions that affect the outcome.

To set up the data, we will need to add variables containing the map coordinates to the data:

```
> d.music.som <- f.ggobi.som(subset(d.music.std,
    select=lvar:lfreq), music.som)
```

Because this data has several useful categorical labels for each row, we will want to keep this information in the data when it is loaded into GGobi:

```
> d.music.som <- data.frame(
  Songs=factor(c(as.character(row.names(d.music)),
    rep("0",36))),
  artist=factor(c(as.character(d.music[,1]),rep("0",36))),
  type=factor(c(as.character(d.music[,2]),rep("0",36))),
  lvar=d.music.som[,1],
  lave=d.music.som[,2],
  lmax=d.music.som[,3],
  lfener=d.music.som[,4], lfreq=d.music.som[,5],
  Map.1=d.music.som[,6], Map.2=d.music.som[,7])
> gd <- ggobi(d.music.som)[1]
```

Add the edges that form the SOM net:

```
> d.music.som.net <- f.ggobi.som.net(music.som)
> edges(gd) <- d.music.som.net + 62
```

And finally color the points according to the type of music:

```
> gcolor <- rep(8,98)
> gcolor[d.music.som$Type=="Rock"] <- 6
> gcolor[d.music.som$Type=="Classical"] <- 4
> gcolor[d.music.som$Type=="New wave"] <- 1
> glyph_color(g) <- gcolor
```

The results can be seen in Figs. 5.9 and 5.10. The plots show two different states of the fitting process and of the SOM net cast through the data. In both fits, a 6 × 6 grid is used and the net is initialized in the direction of the first two principal components. In both fits the variables were standardized to have mean equal to zero and standard deviation equal to 1. The first SOM fit, shown in 5.9, was obtained using the default settings; it gave terrible results. At the left is the map view, in which the fit looks deceptively reasonable. The points are spread evenly through the grid, with rock tracks (purple) at the upper right, classical tracks (green) at the lower left, and new wave tracks (the three black rectangles) in between. The tour view in the same figure, however, shows the fit to be inadequate. The net is a flat rectangle in the 5D space and has not sufficiently wrapped through the data. This is the result of stopping the algorithm too soon, thus failing to let it converge fully.

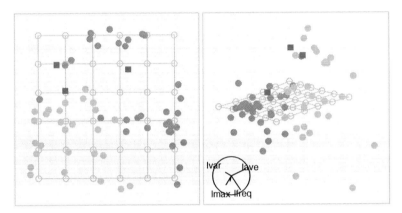

Fig. 5.9. Unsuccessful SOM fit shown in a 2D map view and a tour projection. Although the fit looks good in the map view, the view of the fit in the 5D data space shows that the net has not sufficiently wrapped into the data: The algorithm has not converged fully.

Figure 5.10 shows our favorite fit to the data. The data was standardized, we used a 6×6 net, and we ran the SOM algorithm for 1,000 iterations. The map is at the top left, and it matches the map already shown in Fig. 5.8, except for the small jittering of points in the earlier figure. The other three plots show different projections from the grand tour. The upper right plot shows how the net curves with the nonlinear dependency in the data: The net is warped in some directions to fit the variance pattern. At the bottom right we see that one side of the net collects a long separated cluster of the Abba tracks. We can also see that the net has not been stretched out to the full extent of the range of the data. It is tempting to manually manipulate the net to stretch it in different directions and update the fit.

It turns out that the PCA view of the data more accurately reflects the structure in the data than the map view. The music pieces really are clumped together in the 5D space, and there are a few outliers.

5.3.4 Comparing methods

To compare the results of two methods we commonly compute a confusion table. For example, Table 5.1 is the confusion table for five-cluster solutions

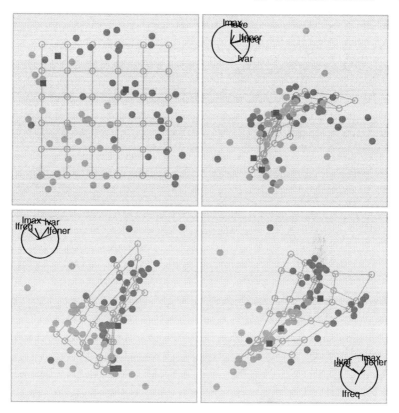

Fig. 5.10. Successful SOM fit shown in a 2D map view and tour projections. Here we see a more successful SOM fit, using standardized data. The net wraps through the nonlinear dependencies in the data, but some outliers remain.

for the Music data from k-means and Ward's linkage hierarchical clustering, generated by:

```
> d.music.dist <- dist(subset(d.music.std,
    select=c(lvar:lfreq)))
> d.music.hc <- hclust(d.music.dist, method="ward")
> cl5 <- cutree(d.music.hc,5)
> d.music.km <- kmeans(subset(d.music.std,
    select=c(lvar:lfreq)), 5)
> table(d.music.km$cluster, cl5)
> d.music.clustcompare <-
    cbind(d.music.std,cl5,d.music.km$cluster)
> names(d.music.clustcompare)[8] <- "Wards"
> names(d.music.clustcompare)[9] <- "km"
> gd <- ggobi(d.music.clustcompare)[1]
```

The numerical labels of clusters are arbitrary, so these can be rearranged to better digest the table. There is a lot of agreement between the two methods: Both methods agree on the cluster for 48 tracks out of 62, or 77% of the time. We want to explore the data space to see where the agreement occurs and where the two methods disagree.

Table 5.1. Tables showing the agreement between two five-cluster solutions for the Music data, showing a lot of agreement between k-means and Ward's linkage hierarchical clustering. The rows have been rearranged to make the table more readable.

k-means	Ward's 1 2 3 4 5		k-means	Ward's 1 2 3 4 5
1	0 0 3 0 14		4	8 2 1 0 0
2	0 0 1 0 0	Rearrange rows $\Rightarrow$	3	0 9 5 0 0
3	0 9 5 0 0		2	0 0 1 0 0
4	8 2 1 0 0		5	0 0 3 16 0
5	0 0 3 16 0		1	0 0 3 0 14

In Fig. 5.11, we link jittered plots of the confusion table for the two clustering methods with 2D tour plots of the data. The first column contains two jittered plots of the confusion table. In the top row of the figure, we have highlighted a group of 14 points that both methods agree form a cluster, painting them as orange triangles. From the plot at the right, we see that this cluster is a closely grouped set of points in the data space. From the tour axes we see that lvar has the largest axis pointing in the direction of the cluster separation, which suggests that music pieces in this cluster are characterized by high values on lvar (variable 3 in the data); that is, they have large variance in frequency. By further investigating which tracks are in this cluster, we can learn that it consists of a mix of tracks by the Beatles ("Penny Lane," "Help," "Yellow Submarine," ...) and the Eels ("Saturday Morning," "Love of the Loveless," ...).

In the bottom row of the figure, we have highlighted a second group of tracks that were clustered together by both methods, painting them again using orange triangles. In the plot to the right, we see that this cluster is closely grouped in the data space. Despite that, this cluster is a bit more difficult to characterize. It is oriented mostly in the negative direction of lave (variable 4), so it would have smaller values on this variable. But this vertical direction in the plot also has large contributions from variables 3 (lvar) and 7 (lfreq). If you label these eight points on your own, you will see that they are all Abba songs ("Dancing Queen," "Waterloo," "Mamma Mia," ...).

We have explored two groups of tracks where the methods agree. In a similar fashion, we could also explore the tracks where the methods disagree.

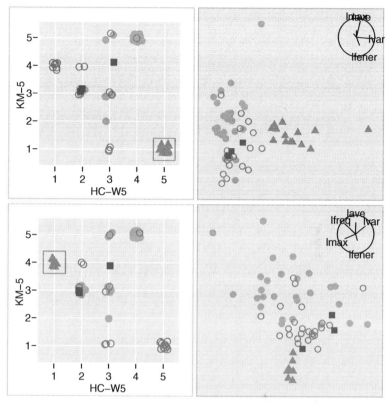

Fig. 5.11. Comparing two five-cluster models of the Music data using confusion tables linked to tour plots. In the confusion tables, k-means cluster identifiers for each plot are plotted against Ward's linkage hierarchical clustering ids. (The values have been jittered.) Two different areas of agreement have been highlighted, and the tour projections show the tightness of each cluster where the methods agree.

5.4 Characterizing clusters

The final step in a cluster analysis is to characterize the clusters. Actually, we have engaged in cluster characterization throughout the examples, because it is an intrinsic part of assessing the results of any cluster analysis. If we cannot detect any numerical or qualitative differences between clusters, then our analysis was not successful, and we start over with a different distance metric or algorithm.

However, once we are satisfied that we have found a set of clusters that can be differentiated from one another, we want to describe them more formally, both quantitatively and qualitatively. We characterize them quantitatively by computing such statistics as cluster means and standard deviations for each variable. We can look at these results in tables and in plots, and we can refine the qualitative descriptions of the clusters we made during the assessment process.

The parallel coordinate plot is often used during this stage. Figure 5.12 shows the parallel coordinate plot for the first of the clusters of music pieces singled out for study in the previous section. Ward's hierarchical linkage and k-means both agreed that these music pieces form a cluster. Since the matrix and the number of clusters are both small, we plot the raw data; for larger problems, we might plot cluster statistics as well [see, for example, Dasu, Swayne & Poole (2005)].

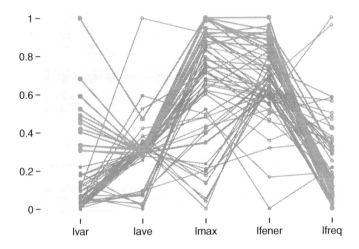

Fig. 5.12. Characterizing clusters in a parallel coordinate plot. The highlighted profiles correspond to one cluster for which Ward's hierarchical linkage and k-means were in agreement.

This cluster containing a mix of Beatles and Eels music has high values on lvar, medium values of lave, high values of lmax, high values of lfener, and varied lfreq values. That is, these pieces of music have a large variance in frequency, high frequency, and high energy relative to the other music pieces.

5.5 Recap

Graphics are invaluable for cluster analysis, whether they are used to find clusters or to interpret and evaluate the results of a cluster analysis arrived at by other means.

The spin and brush approach can be used to get an initial look at the data and to find clusters, and occasionally, it is sufficient. When the clustering is the result of an algorithm, a very useful first step is to paint the points by cluster membership and to look at the data to see whether the clustering seems sensible. How many clusters are there, and how big are they? What shape are they, and do they overlap one another? Which variables have contributed most to the clustering? Can the clusters be qualitatively described? All the plots we have described can be useful: scatterplots, parallel coordinate plots, and area plots, as well as static plots like dendrograms.

When the clusters have been generated by a model, we should also use graphics to help us assess the model. If the model makes distributional assumptions, we can generate ellipses and compare them with the clusters to see whether the shapes are consistent. For self-organizing maps the tour can assist in uncovering problems with the fit, such as when the map wraps in on itself through the data making it appear that some cases are far apart when they are truly close together. A confusion table can come alive with linked brushing, so that mismatches and agreements between methods can be explored.

Exercises

1. Using the spin and brush method, uncover three clusters in the Flea Beetles data and confirm that these correspond to the three species. (Hint: Transform the data to principal components and enter these variables into the projection pursuit guided tour running the holes index. Also the species should not be identified by color or symbol, until the clusters have been uncovered.)
2. Run hierarchical clustering with average linkage on the Flea Beetles data (excluding species).
    a) Cut the tree at three clusters and append a cluster id to the dataset. How well do the clusters correspond to the species? (Plot cluster id vs species, and use jittering if necessary.) Using brushing in a plot of cluster id linked to a tour plot of the six variables, examine the beetles that are misclassified.
    b) Now cut the tree at four clusters, and repeat the last part.
    c) Which is the better solution, three or four clusters? Why?
3. For the Italian Olive Oils,
    a) Consider the oils from the four areas of Southern Italy. What would you expect to be the result of model-based clustering on the eight fatty acid variables?
    b) Run model-based clustering on the Southern oils, with the goal of extracting clusters corresponding to the four areas. What is the best model? Create ellipsoids corresponding to the model and examine these in a tour. Do they match your expectations?
    c) Create ellipsoids corresponding to alternative models and use these to decide on a best solution.

4. This question uses the Rat Gene Expression data.
   a) Explore the patterns in expression level for the functional classes. Can you characterize the expression patterns for each class?
   b) How well do the cluster analysis results match the functional classes? Where do they differ?
   c) Could you use the cluster analysis results to refine the classification of genes into functional classes? How would you do this?
5. In the Music data, make further comparisons of the five-cluster solutions of k-means and Ward's hierarchical clustering.
   a) On what tracks do the methods disagree?
   b) Which track does k-means consider to be a singleton cluster, while Ward's hierarchical clustering groups it with 12 other tracks?
   c) Identify and characterize the tracks in the four clusters where both methods agree.
6. In the Music data, fit a 5 × 5 grid SOM, and observe the results for 100, 200, 500, and 1,000 updates. How does the net change with the increasing number of updates?
7. There is a mystery dataset in the collection, called clusters-unknown.csv. How many clusters are in this dataset?

6
Miscellaneous Topics

Analysts often encounter data that cannot be fully analyzed using the methods presented in the preceding chapters. In this chapter we introduce, however briefly, some of these different kinds of data and the additional methods needed to analyze them, beginning with a section on inference. A more complete treatment of each topic will be found on the book web site.

6.1 Inference

Revealing, informative plots often provoke the question "Is what we see really there?" When we see a pattern in a plot, it is sometimes completely clear that what we see is "really there," and that it corresponds to structure in the data, but sometimes we are not so sure.

What is meant by *really there*? Sometimes a pattern is detected, either numerically or visually, even when there is nothing happening. In the hypothesis testing context, we might say that the pattern is consistent with a null scenario. For example, in a simple linear regression scenario with two variables X, Y, we are interested in the *dependence* between the two variables. We call the presence of this dependence the *alternative hypothesis*, and then the natural *null hypothesis* is that the two variables are *independent*. A common numerical "pattern" that suggests linear dependence is the correlation. The plots in Fig. 6.1 all show pairs of variables with correlation close to 0.7, but most of these plots are not consistent with the null hypothesis. The top left plot is the only one that shows the relationship we probably imagine when we hear that two variables have a correlation of 0.7, a perfect example of linear dependence between X and Y.

The remaining plots show very different departures from independence. The top right plot shows an example where the correlation is misleading: The apparent dependence is due solely to one sample point, and the two variables are in fact not dependent. The plot at bottom left shows two variables that are dependent, but the dependence is among sub-groups in the sample and it

130   6 Miscellaneous Topics

is negative rather than positive. The plot at bottom right shows two variables with some positive linear dependence, but the obvious non-linear dependence is more interesting.

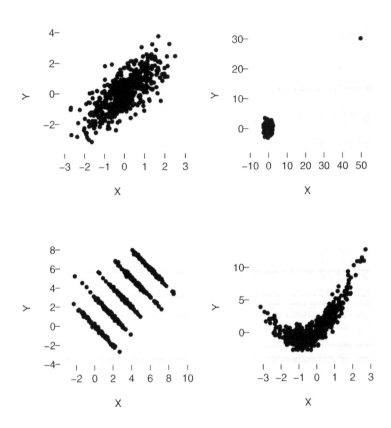

Fig. 6.1. Studying dependence between X and Y. All four pairs of variables have correlation approximately equal to 0.7, but they all have very different patterns. Only the top left plot shows two variables matching a dependence modeled by correlation.

With graphics we cannot only detect a linear trend, but virtually any other trend (nonlinear, decreasing, discontinuous, outliers) as well. That is, we can easily detect many different types of dependence with visual methods.

The first step in using visual methods to determine whether a pattern is "really there" is to identify an appropriate pair of hypotheses, the null and an alternative. The second step is to determine a process that simulates the null hypothesis to generate comparison plots. Some common null hypothesis scenarios are as follows:

6.1 Inference   131

1. Distributional assumptions are tested by simulating samples from the distribution using parameters estimated from the sample. For example, if we suspect, or hope, that the data is consistent with a normal population, we simulate samples from a normal distribution using the sample mean and variance–covariance matrix as the parameters. This approach was used in Sect. 4.1.1, because linear discriminant analysis assumes that the data arises from a multivariate normal mixture.
2. Independence assumptions are tested using permutation methods, where the appropriate columns of data values are shuffled to break the association in the ordered tuples. When there are two variables, shuffle one of the columns, say the X-values, and make plots of the permuted data to compare with the real data. This approach arises from the methods used in permutation tests (Good 2005).
3. In labeled data problems, the assumption that the labels matter is tested by shuffling the labels. For example, in a designed experiment with two groups, control and treatment, randomly re-assign the control and treatment labels. Similarly, for a supervised classification problem, shuffle the class ids.

Under these scenarios the *test statistic* is the plot of the real data. If the plot of the data is noticeably different from all of the plots of data generated under the null scenario, then we have evidence for the presence of real structure in the data.

For an example of making and testing inferences in a supervised classification problem, we use the Flea Beetles data. Our goal is to find clusters corresponding to the three different species:

```
> library(rggobi)
> d.flea <- read.csv("flea.csv")
> attach(d.flea)
> gd <- ggobi(d.flea)[1]
```

Once the points have been colored according to species, a projection pursuit guided tour using the LDA index is used to find clusters corresponding to the species:

```
> ispecies = as.integer(species)
> gtype <- rep(6,74)
> gtype[ispecies==1] <- 4; gtype[ispecies==3] <- 3
> glyph_type(gd) <- gtype; glyph_size(g) <- 3

> gcolor = rep(1,74)
> gcolor[ispecies==1] <- 5; gcolor[ispecies==3] <- 6
> glyph_color(gd) <- gcolor
```

Repeat the process numerous times with the species labels scrambled:

132   6 Miscellaneous Topics

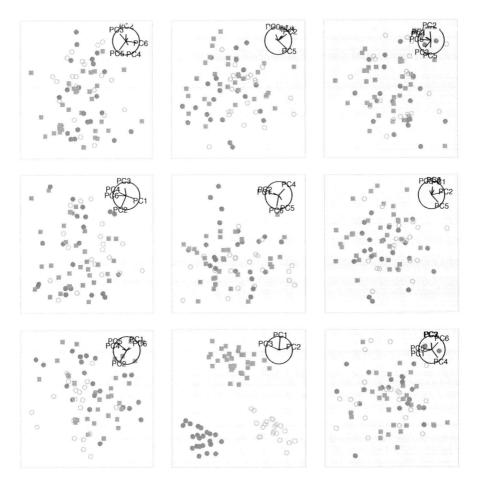

Fig. 6.2. Testing the assumption that species labels correspond to three groups in the data, using a projection pursuit guided tour with the LDA index. The plots show optional projections of the same data, except that in eight of the nine plots, the species labels have been permuted. The plot of the real data is the odd one out, and very easy to spot!

```
> pspecies <- sample(ispecies)

> gtype <- rep(6,74)
> gtype[pspecies==1] <- 4; gtype[pspecies==3] <- 3
> glyph_type(gd) <- gtype

> gcolor <- rep(1,74)
> gcolor[pspecies==1] <- 5; gcolor[pspecies==3] <- 6
> glyph_color(gd) <- gcolor
```

In Fig. 6.2, the plot of the best projection of the real data is embedded among plots of the best projections of the data with permuted labels. Which one is different? The real data plot is easily distinguished from the rest, offering evidence for the existence that the data space includes clusters corresponding to the labeled classes.

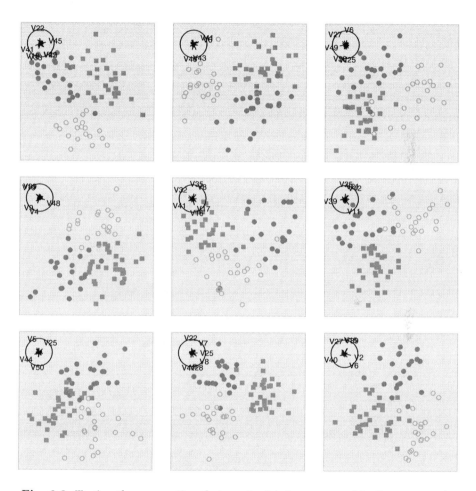

Fig. 6.3. Testing the assumption that species labels correspond to three groups in simulated data, using a projection pursuit guided tour with the LDA index. The data is similar to Flea Beetles in that there are 74 cases, divided into three classes, but there are now 50 variables instead of 6. The plots show optional projections of the same data, except that in eight of the nine plots, the species labels have been permuted. The plot of the real data (left plot of the middle row) is not distinguishable from the others!

In contrast, Fig. 6.3 shows an example where the data is consistent with the null hypothesis. The data used here is based on the Flea example, except that here we have created 50 new variables, each samples of size 74 from a univariate standard normal distribution. To create the "real" data, the species labels are the same as the Flea Beetles data, and to create the comparison datasets, these labels are permuted. There should be no difference between the real data and the permuted label data, because there are no real clusters in the data. The plots verify this. Each is a projection produced by a projection pursuit guided tour using the LDA index. The real data (left plot of middle row) shows clustering of the three species, but so do the other plots, produced by permuting the species labels. The real data is consistent with the null hypothesis that there is no real clustering of the data. The apparent clustering of the data occurs because we have a small number of data points in high dimensions. Here is the code that produced the simulated data:

```
> x <- matrix(rnorm(74*50),ncol=50)
> colnames(x) <- paste("V",1:100,sep="")
```

The remaining code to color the points (first by species, and again after permuting the species labels) is the same as that given on the preceding pages for the Flea Beetles data.

In summary, it is possible to build inference into graphical exploration of data. The full chapter available on the web provides more background and several more examples.

6.2 Longitudinal data

In longitudinal data (sometimes called panel data), individuals are repeatedly measured through time, enabling the direct study of change (Diggle, Heagerty, Liang & Zeger 2002). A longitudinal study focuses on the way the relationship between the response variables and the covariates changes. Unique to longitudinal studies is the ability to study individual responses. In repeated cross-sectional studies, on the other hand, the study of individual responses is not supported, because a new sample is taken at each measurement time; this supports studies of societal trends but not individual experiences. A longitudinal dataset supports such investigations because it is composed of a collection of time series, one for each individual. The challenge posed by the data arises because these time series may not be uniform in structure. Each individual may have special characteristics, and measurements on different topics or variables may be taken each time an individual is measured. The time characteristics of the data can also vary between individuals, in the number and value of the time points, and in the intervals between them.

These irregularities yield data that is difficult to work with. If it were equispaced over time, and if each individual had the same number of time points and covariates, it would be easier to develop formal models to summarize trends and covariance. Still, if we want to learn what this data has to offer, we have to find a way to work with it.

To organize the data, we denote the *response* variables to be Y_{ijt_i}, and the time-dependent explanatory variables, or *covariates* to be X_{ikt_i}, where $i = 1, \ldots, n$ indexes the number of individuals in the study, $j = 1, \ldots, q$ indexes the number of response variables, $k = 1, \ldots, p$ indexes the number of covariates, and $t_i = 1, \ldots, n_i$ indexes the number of times individual i was measured. Note that n is the number of subjects or individuals in the study, n_i is the number of time points measured for individual i, q is the number of response variables, and p is the number of explanatory variables, measured for each individual and time. The explanatory variables may include indicator variables marking special events affecting an individual. There may also be time-independent explanatory variables or covariates, which we will denote as Z_{il}, $i = 1, \ldots, n$, $l = 1, \ldots, r$.

For the Wages data, there are $n = 888$ individuals, $q = 1$ response variable (lnw), and $p = 1$ time-dependent covariate (uerate). The measurements are time-indexed by the length of time in the workforce (exper), measured in years, when the wages were recorded. Note that the data does not include any variable representing time per se, but that exper stands in for time. We reorganize the data into two tables. The first table contains the time variable and the time-dependent measurements: $Y_{ij}\text{exper}_i$, $X_{ik}\text{exper}_i$ (exper, lnw, uerate). The second table, much shorter, contains the $r = 4$ time-independent covariates (ged, black, hispanic, and hgc). In addition, both tables include the subject id's, and that is how they are linked. (Portions of the tables are shown in Chap. 7.)

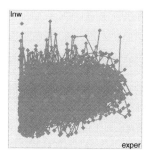 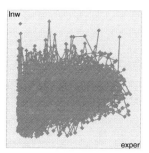 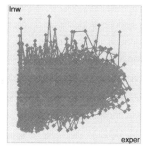

Fig. 6.4. Looking over the longitudinal Wages data. Longitudinal plots of sampled subjects, highlighted in orange against all the data, showing variability from subject to subject.

With so many subjects, an initial plot of lnw against exper is nothing but a tangle of over-plotted lines. Imagine looking at the plots in Fig. 6.4 without any highlighting!

136   6 Miscellaneous Topics

```
> library(rggobi)
> gg <- ggobi() # Open wages3.xml, and draw the edges
> d.wages1 <- gg[1]
> d.wages3 <- gg[3]
```

To get acquainted with any one subject, we can leave the plot in place and highlight the profile for that subject in a contrasting color. A quick overview of the variability from subject to subject can be obtained by animating that process for a sample of subjects, highlighting their profiles one by one:

```
> smp <- sample(1:888,50)
> for (i in smp) {
    gtype <- rep(1,6402); gcolor <- rep(1,6402)
    gtype[d.wages1[[1]]==d.wages3[[1]][i]] <- 6
    gcolor[d.wages1[[1]]==d.wages3[[1]][i]] <- 9
    glyph_type(d.wages1) <- gtype
    glyph_color(d.wages1) <- gcolor
    glyph_size(d.wages1) <- 4
  }
```

The profiles of three subjects are highlighted in Fig. 6.4. The subject highlighted in the left plot has a slow but steady increase in wages as his experience grows, suddenly terminated by a big drop. For the second subject, we have only a few early measurements, during which time his wages increased rapidly. For the third subject, we have once again only a few measurements capturing the beginning of his work life, but his wages fluctuated wildly. Using the animation to study more subjects, we are surprised to see how much the wage profiles differ from subject to subject!

This variability suggests that we might try to summarize the data by characterizing the different types of profiles. We will calculate several numerical indicators for each profile and use these to study the subjects. Initially, we count the number of data points for each subject so that we can exclude subjects with very few measurements.

```
> wages.count <- summary(d.wages1[[1]], maxsum=888)
```

We then calculate descriptors for the profiles: standard deviation of lnw (log wages), standard deviation of the differences between consecutive measurements, and linear trend, calculated both classically and robustly.

6.2 Longitudinal data

```
> wages.linear <- rep(0,888)
> wages.rlin <- rep(0,888)
> wages.sd <- rep(0,888)
> wages.sddif <- rep(0,888)
> for (i in 1:888) {
+   x <- d.wages1[d.wages1[[1]]==d.wages3[[1]][i]][3:2]
+   if (dim(x)[1]>1) {
+     wages.sd[i] <- sd(x)
+     difs <- NULL
+     for (j in 2:length(x))
+       difs <- c(difs,
+         d.wages1[[2]][x[j]] - d.wages1[[2]][x[j]-1])
+     wages.sddif[i] <- sd(difs)
+   }
+   if (dim(x)[1]>2)
+     wages.linear[i] <- coef(lm(lnw~exper,data=x))[2]
+   if (dim(x)[1]>3) {
+     wgts <- 1/(residuals(lm(lnw~exper,data=x))^10+1)
+     wages.rlin[i] <- coef(lm(lnw~exper,data=x,
+       weights=wgts))[2]
+   }
+   cat(i," ")
+ }
```

Add this additional data to GGobi to explore profiles with different characteristics:

```
> gg["descriptors"] <- data.frame(id=d.wages3$id,
    count=wages.count, sd=wages.sd,
    sddif=wages.sddif, linear=wages.linear,
    rlin=wages.rlin)
```

Figure 6.5 illustrates exploring profiles by linked brushing in plots of these descriptors. (Subjects with only one or two measurements have been excluded from these plots.) In the top two rows of plots we examine the volatility in wage experiences by plotting the standard deviations of all the subjects' measurements (sd) against the standard deviations of differences between consecutive measurements (sddif). Large values of both indicate subjects who have a more varied wage experience (top row). Large values of sd but small values of sddif indicate subjects whose wage experience has varied over time but in a consistent manner, such as a steady increase or decrease (middle row). In the bottom row of plots we examine the slope from a robust linear fit (rlin) relative to sddif to find subjects whose wages have steadily increased.

In summary, it is possible to penetrate difficult data like ragged time longitudinal data using interactive graphics. From this data, we have already learned that the men in this study have widely varied wage experiences. We

138 6 Miscellaneous Topics

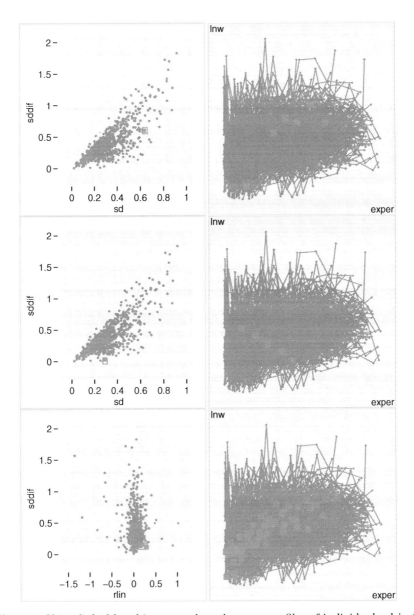

Fig. 6.5. Using linked brushing to explore the wage profiles of individual subjects.

would probably clean the data further before doing a more detailed analysis, removing subjects with too few measurements or too short a time in the workforce. The full chapter develops this analysis further and offers more approaches to longitudinal data.

6.3 Network data

Fundamental to network data is a graph. A graph consists of a set of points (also called nodes or vertices), some of which are connected by edges, and this common mathematical object appears in many areas of computer science, engineering, and data analysis. In telecommunications, for example, a node represents a network element, an IP address, or a telephone number, and an edge represents the network connection or traffic between a pair of nodes. In *social network analysis* (Wasserman & Faust 1994), a node most often represents a person, and an edge the connections between two people. Social network analysis emerged as an academic discipline in about the 1930s, with networks studied for a wide variety of purposes: to analyze organizational behavior; to understand the process by which a disease is spread; to identify suspected criminals; and to study social taboos in unfamiliar cultures. This discipline has recently become quite well known because of the popularity of Facebook® and MySpace®, web sites where people register (adding a new node in the graph) and then create links to their friends (adding edges to the graph).

In analyzing network data, we are interested in the attributes of the graph itself as well as those of the nodes and edges of which it is composed.

For example, a graph may be *fully connected*, in which every node is connected to every other node, or it may be *sparse*, with few edges. A graph may show clusters, distinct *sub-graphs* that are densely connected within themselves but very sparsely connected to one another. Such patterns correspond to sub-communities that interact only through a few influential individuals.

That presents an example of an interesting node attribute, because we may want to identify these influential individuals. One way to measure influence is the *degree* of a node, which is the number of other nodes to which it is directly connected. Yet another is the *betweenness centrality*, a measure of a node's influence based on the number of minimum paths that pass through it. Many of these attributes can be identified algorithmically, and there is a large body of work in discrete mathematics and computer science on which these algorithms are based.

Edges, of course, have interesting attributes as well, such as weights. In *directed graphs*, edge direction is significant, and then the edges are most often represented by arrows. An example is a social network in which a person provides a list of people they consider friends, establishing a set of edges from the subject's node to the nodes of their friends.

6 Miscellaneous Topics

Graph layout algorithms read the lists of nodes and edges, and generate a position for each node with the goal of revealing structure in the graph [see, for example, Di Battista, Eades, Tamassia & Tollis (1998)]. A good layout algorithm will also use data ink efficiently and avoid distracting artifacts. The algorithm will manage properties such as edge length, edge crossings, number of different slopes, node separation, and aspect ratio. There is usually no single best layout for any graph, and it is useful to look at the same graph in several different layouts, since each one may reveal different global or local structure in the graph. There are many commercial and open source software packages designed for graph layout. Galleries of sample graphs on graphviz.org, aisee.com, and tomsawyer.com demonstrate the variety of graphs and graph layout algorithms.

For a statistician studying graph data, the layout and description of the graph may be only part of the story, because the nodes and edges may each have a number of attributes, both real-valued and categorical. We may know the gender, age, or other demographic variables for each person in a social network; if we are conducting a study of disease transmission, we even may have longitudinal data representing repeated medical test results. Edge data may also be as simple as a single number, such as a real-valued measure of frequency of contact, or as complex as a multivariate time series capturing contacts of many types.

The richer the node and edge data becomes, the more valuable it is to be able to display the graph in the context of interactive data visualization software that allows us to link the graph to graphical displays of the associated data.

The following example is based on the Florentine Families data. The data comes from a study of the families of Renaissance Florence, where connections between about 50,000 people from 1,300 to 1,500 were constructed from primary sources (Padgett & Ansell 1993). A tiny subset of this data that contains information about 16 families is used here — it is a canonical dataset in social network analysis.

```
> library(graph, SNAData)      # Bioconductor packages
> data(florentineAttrs, business, marital)
> families = florentineAttrs   # nodes
> ties = families[,"NumberTies"]
> gg = ggobi(families)
```

The families will be the nodes in the graph. For these families, we have three real-valued variables: their wealth in 1,427, their total number of business and family ties within a larger set of families, and their number of seats on the governing body of Florence (called Priorates). In addition, there are two sets of edges. In one set, a link between two families means that they have financial dealings with one another:

```
> e = t(edgeMatrix(business)) # first set of edges
```

```
> # Prepare to add edge names
> src = nodes(business)[e[,1]]
> dest = nodes(business)[e[,2]]
> edgenames = paste(src, dest, sep="->")
> # Add edge weights: the average of the two node variables
> weights = matrix((ties[e[,1]] + ties[e[,2]]) / 2, ncol=1)
> dimnames(weights) = list(edgenames, c("meanTies"))
>
> gg$business = data.frame(weights)
> edges(gg$business) = cbind(src, dest)
```

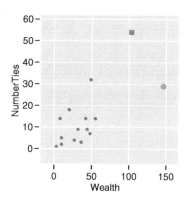

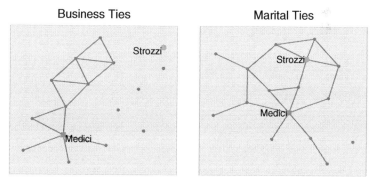

Fig. 6.6. Business and marital relationships among 16 Renaissance Florentine families.

We also added an edge attribute. A weight for each edge was calculated by averaging the NumberTies of the source and destination nodes. An ideal edge weight would convey the number of business ties between the two families, but the dataset does not include that information, so instead we are using a weight that is large when the average number of *all* business ties for the

142   6 Miscellaneous Topics

two connected families is high. We can use this to draw our attention to links between well-connected families.

The other edge set contains marital relationships, and we use similar code to add this second set of edges. We then have two distinct graphs with a common set of nodes but different edges. How different are they?

In order to look at the graphs, we have to lay them out. We use the neato layout algorithm from the graphviz(Gansner & North 2000) library; it is available from GGobi if the library is installed.

The bottom row of plots in Fig. 6.6 shows the two graphs, and they are quite different. Most families are connected by marriage; only one family is not connected, as shown by the single unconnected point in the graph of marital ties. In contrast, five families do not have financial dealings with other families in the data, as shown by the five unconnected points in the graph of business ties.

We might also identify the dominant families, using other information about each family. In a scatterplot of NumberTies versus Wealth, at the top of 6.6, we highlight the Medicis with an orange square and the Strozzis with a green circle. Both families have a lot of ties with other families and great wealth. From the graphs, we see that they seem to have played very different roles in the society. Both families were highly interconnected by marriage to other families — although not directly to each other — but only the Medicis had many financial relationships.

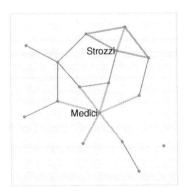 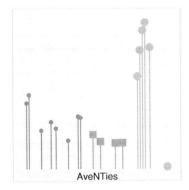

Fig. 6.7. Highlighting the links involving the greatest number of ties. The average shifted histogram (ASH) **(right)** plots the edge weights, and the points in the ASH are linked by color to the edges in the graph.

Next we study the connections between families. Plots of the edge weight variable, AveNTies, can be linked to the graphs. Figure 6.7 shows an ASH of AveNTies linked to the graph of marital relationships. We have highlighted in green the links with the highest values of AveNTies, which reflect both marital and business ties. We see that most links are connected to the Medici node, which leads us to believe that the large edge weights reflect the many financial activities of the Medici family. The edges with the next highest values of AveNTies are all connected to the Strozzi node; as we know, this family's relationships are principally through marriage.

Even though the marital graph does not show clustering in any obvious way, this pair of graphs reinforces the impression of a division within the dominant families of Florence of that era.

Graphs can also exist in higher dimensions. We use an adjacent transposition graph as an example. This family of graphs has special meaning to discrete mathematicians, and Thompson (1993) has also used it to depict relationships between surveyed preference judgments, such as the results of an election.

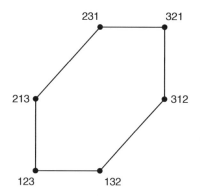

Fig. 6.8. The $n = 3$ adjacency transposition graph.

The $n = 3$ adjacent transposition graph is generated as follows. We start with all permutations of the sequence 1, 2, 3. There are 3!, or six, such sequences, and we make each a vertex in the graph. We connect two vertices by an edge if one permutation can be turned into the other by simply transposing two adjacent elements. This graph is easy to lay out and draw by hand, because it is nothing more than a hexagon, as shown in Fig. 6.8. Notice that the $n = 3$ graph shows no edge crossings in the plane, the two-dimensional (2D) layout space.

Now consider the $n = 4$ adjacent transposition graph. This graph has $n! = 24$ vertices. The rule for connecting vertices works such that the node corresponding to 2134 is connected to the node for 1234 because one can be turned into the other by swapping the middle two digits. However, 2134 is not connected to 1423 because there is no way to move from one to the other with a single adjacent flip. How many flips would it take? What we would like is a representation of the graph that makes that easier to see, and this is not something most of us can draw by hand.

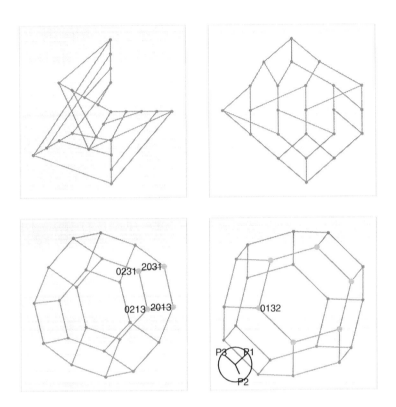

Fig. 6.9. The $n = 4$ adjacency transposition graph in various layouts: `radial`, `dot`, `neato` in the plane, and `neato` in three dimensions (3D). After the 3D layout, we see that the resulting object is composed of squares and hexagons.

We will experiment with the different graph layout algorithms. Figure 6.9 shows four different layouts for this graph. Most layout methods used in GGobi are available through `graphviz`; the exception is the `radial` layout algorithm, which was implemented based on Wills (1999).

The first layout reveals a very structured object, but it includes many edge crossings and offers no illumination about the structure of the data. The algorithm used (`radial`) takes the first point as its center and lays out the rest of the points in concentric circles around it. It is obviously inappropriate for this data, but it can be very helpful with some kinds of social network data.

The second layout is even more regular, but it is not right either. It was produced using `dot`, which is a hierarchical layout algorithm that is designed for trees. Since we know this graph far from being a tree, we just tried out this algorithm for fun.

With the third and fourth layouts, we hit pay dirt. They were both produced with neato, which is a layout algorithm designed for just this kind of data: *undirected* graphs, where it makes no difference whether we say that an edge is from 1243 to 1234 or from 1234 to 1243. Neato constructs a virtual physical model, as if each edge were a spring, and runs an iterative solver to minimize the energy of the model. The layout shown on the left was produced in 2D, and the one on the right in 3D. They do not look much different here, but after producing the 3D layout, we can manipulate it in a 2D tour and look at it from all angles. When we do that, the edge crossings that interfere with interpretation in the 2D projection are not a problem.

What do we see? The resulting shape is formed of hexagons and squares; one of each is highlighted in the two neato plots. The shape is a *truncated octahedron*. Its component shapes, in the language of graph theory, are *six cycles* and *four cycles*, i.e. "loops" composed of six and four points. We leave it to you to look at the graph in GGobi and see what characterizes these structures (Littman, Swayne, Dean & Buja 1992).

In summary, network data analysis requires methods and algorithms that illuminate the structure in the graph itself, but it may also require the full toolkit used for multivariate data analysis. Even when only limited data is available for the nodes and edges, laying out a graph in a high-dimensional space can make it easier to interpret complex structures.

6.4 Multidimensional scaling

Multidimensional scaling (MDS) is a family of methods for studying the similarity between objects, in particular for finding a low-dimensional representation where more similar objects are close to each other. For example, researchers in the social sciences may study the similarity in the ratings for pairs of stimuli such as tastes, colors, and sounds. In marketing studies the pairs may be brands or products, such as soft drinks or breakfast cereals. MDS can also be used for graph layout, which was introduced in the preceding section. The number of edges in the shortest path between each pair of nodes is used to derive a set of distances that can be subjected to MDS.

MDS begins with proximity data, which consists of similarity (or dissimilarity) information for pairs of objects. If the objects are labeled $1, ..., n$, the proximity data is described by dissimilarity values $D_{i,j}, i, j = 1, ..., n$: Pairs of similar objects will have small values of $D_{i,j}$, and pairs of dissimilar objects will have larger values. If the data is provided as similarities, it can be converted to dissimilarities by some monotone decreasing transformation. The goal of MDS is to map the n objects to points $x_1, ..., x_n$ in some lower-dimensional space in such a way that the $D_{i,j}$ are well approximated by the distances $||x_i - x_j||$ in the reduced dimension, so as to minimize a *stress function*, for example:

$$\text{Stress}_D(x_1, ..., x_n) = \left(\sum_{i \neq j = 1...n} (D_{i,j} - ||x_i - x_j||)^2 \right)^{1/2}.$$

The outer square root is just a convenience that gives greater spread to small values. The minimization can be carried out by straightforward gradient descent applied to Stress_D, viewed as a function on R^{kn}. This formulation of MDS is from Kruskal (1964a) and Kruskal (1964b), which refined the original classical scaling proposed by Torgerson (1952).

The data MDS is applied to is therefore a bit different than the input data for most multivariate methods. In a more typical multivariate problem, the analyst starts with an $n \times p$ data matrix, where the rows correspond to the n objects, or samples, and the columns correspond to the p variables recorded for each object. We may sometimes calculate an $n \times n$ interpoint distance matrix, for example, as a step in cluster analysis. This $n \times n$ distance matrix is exactly the starting point for MDS.

The distance matrix in Fig. 6.10 corresponds to the simple 5×5 grid pictured. The distances are obviously not Euclidean; rather, they represent the number of edges in the shortest path between each pair of nodes.

$$\mathbf{D} = \begin{bmatrix} 0 & 1 & 2 & 3 & 4 \\ 1 & 0 & 3 & 4 & 5 \\ 2 & 3 & 0 & 5 & 6 \\ 3 & 4 & 5 & 0 & 7 \\ 4 & 5 & 6 & 7 & 0 \end{bmatrix}$$

Fig. 6.10. A 5×5 grid and the corresponding distance matrix, with entries representing the number of edges in the shortest path between pairs of nodes.

Many MDS methods exist, and the stress function is the main source of difference among them. Here are two dichotomies that allow us to structure some possibilities:

6.4 Multidimensional scaling

- Kruskal–Shepard distance scaling versus classical Torgerson–Gower inner-product scaling: Distance scaling is based on direct fitting of distances to dissimilarities, whereas the older classical scaling is based on a detour whereby dissimilarities are converted to a form that is naturally fitted by inner products $< xi, xj >$.
- Metric scaling versus nonmetric scaling: Metric scaling uses the actual values of the dissimilarities, whereas nonmetric scaling effectively uses only their ranks (Shepard 1962, Kruskal 1964a). Nonmetric MDS is realized by estimating an optimal monotone transformation $f(D_{i,j})$ of the proximities simultaneously with the configuration.

A conceptual difference between classical and distance scaling is that inner products rely on an origin, whereas distances do not; a set of inner products determines uniquely a set of distances, but a set of distances determines a set of inner products only modulo change of origin. To avoid arbitrariness, one constrains classical scaling to configurations with the mean at the origin.

The computational difference between classical and distance scaling is that the minimization problem for classical scaling can be solved with a simple eigen-decomposition, whereas distance scaling requires iterative minimization.

Incorporating MDS into dynamic and interactive data visualization software offers significant advantages over using it in a static context. In addition to all the tools described in earlier chapters, such as touring and brushing, some advantages are specific to the embedding of an iterative algorithm in a dynamic system. A tool in GGobi, ggvis, supports interactive MDS. It is a direct descendant of XGvis (Littman et al. 1992, Buja & Swayne 2002) and has much in common with another earlier system, ViSta–MDS (McFarlane & Young 1994). These systems provide two important interactive capabilities:

1. Animated optimization: The configuration points are displayed continuously as they are subjected to MDS optimization. At the same time, the values of the cost function are also shown in a trace plot.
2. Manual dragging: Configuration points can be moved interactively with mouse dragging.

Figure 6.11 shows nine "snapshots" taken during an MDS animation. We performed a Stress minimization in three dimensions to the 5×5 grid. This provides a flavor of the use of MDS in graph layout, because this simple grid is actually a highly regular graph, with 25 nodes and 40 edges. The final pictures show some curvature, which is a result of the non-Euclidean distances.

For an example of applying MDS to data, we turn to the well-known Rothkopf **Morse Code Confusion Rates** data (Rothkopf 1957), which is to MDS what Fisher's Iris data is to discriminant analysis. Morse codes represent each letter and digit with a sequence of between two and five characters, a combination of dots and dashes. This data originated in an experiment where inexperienced subjects were exposed to pairs of Morse codes in rapid order.

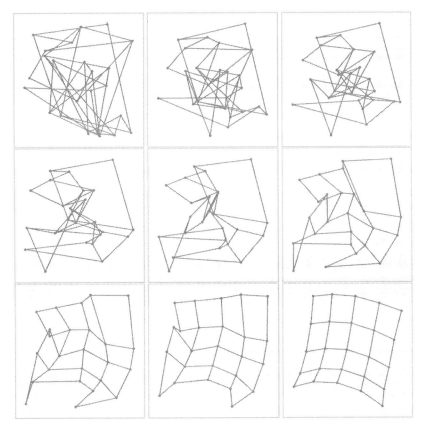

Fig. 6.11. Snapshots showing nine stages of a Stress minimization of a 5 × 5 grid in three dimensions.

The subjects had to decide whether the two codes in a pair were identical. The results were summarized in a table of confusion rates.

Confusion rates are similarity measures: Codes that are often confused are interpreted as similar or close. Similarities, $S_{i,j}, i,j = 1,...,n$ were converted to dissimilarities using

$$D_{i,j}^2 = S_{i,i} + S_{j,j} - 2S_{i,j}$$

This function also symmetrizes the dissimilarities, $D_{i,j} = D_{j,i}$, which is needed by MDS.

Applying metric Kruskal–Shepard distance scaling in $k = 2$ dimensions to the Morse code dissimilarities produced the configuration shown in Fig. 6.12. Generally, codes of the same length are close to each other, as are codes with the same fraction of dots. The length of codes roughly increases from left to right, and the fraction of dots decreases from bottom to top. [These

6.4 Multidimensional scaling

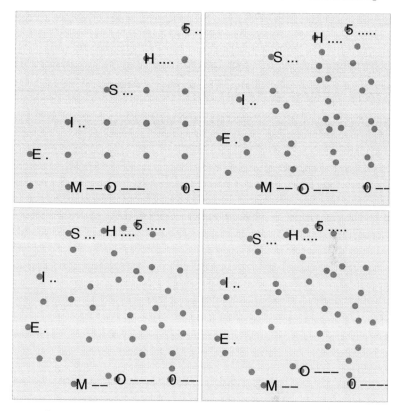

Fig. 6.12. Applying Kruskal–Shepard metric MDS scaling to the Morse Code confusion data: **(top left)** initial configuration, **(bottom right)** final configuration. Generally, codes of the same length are close to each other, as are codes with the same fraction of dots. The length of codes roughly increases from left to right, and the fraction of dots decreases from bottom to top.

observations agree with the many published accounts (Shepard 1962, Kruskal & Wish 1978, Borg & Groenen 2005).] The striping or banding of points in the configuration corresponds to strips of codes of the same length but different combinations of dots and dashes. Drawing edges to connect codes of the same length makes this pattern clearer (Fig. 6.13).

Choosing a different scaling metric, or different parameters, will lead to a different layout. The right plot in Fig. 6.13 shows the result for Kruskal–Shepard non-metric MDS. Many other parameters to the algorithm can be changed, and these as with the choice of the initial configuration will affect the final result.

MDS can also create configurations in dimensions other than 2. For the Morse code data, a layout in 3D is revealing: Points are arranged approximately on a sphere. Codes of length 1 are close to each other but far from

150   6 Miscellaneous Topics

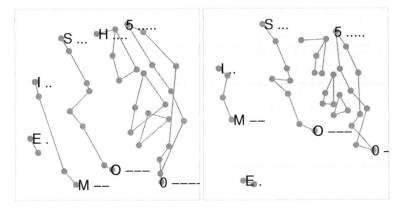

Fig. 6.13. Studying the results of MDS. Edges connecting code of the same length are added, making it easier to see strips in the layout corresponding to the codes of the same length. At right are the results of non-metric MDS.

other length codes, and the same can be said for codes of length 2. With identification we can also learn that points corresponding to the letters K (-.-) and D (-..) are closer to some of the length four codes, such as the letter L (.-..), than to other length 3 codes.

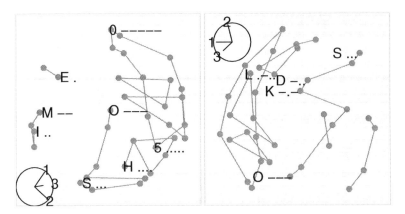

Fig. 6.14. Layout of the Morse Code in 3D is revealing. The points lie approximately on a sphere.

More information about MDS metrics and the ggvis parameters is available in Buja, Swayne, Littman, Dean & Hofmann (2007), from which this section has largely been excerpted.

Exercises

1. For the Music data, permute the Type and search for the best separation. Is the separation as good as the separation for the true class variable?
2. Subset the Australian crabs data, choosing only the blue males. Compute the mean and variance–covariance of this subset, and use this information to generate a sample from a multivariate normal distribution having the same population mean and variance. Compare this simulated data with the actual data. Do you think the blue male crabs subset is consistent with a sample from a multivariate normal? Justify your opinion.
3. For the Wages data:
   a) Identify a person whose wages steadily increase with experience.
   b) Identify a person whose wages are very volatile throughout their experience.
   c) Compute the maximum minus minimum wage for each person, and use this to plot the profiles of 10 people having the widest range in wages.
   d) Compute the correlation between uerate and lnw for each person. Use this to find the person with the strongest relationship between unemployment rate and wages. Plot the profiles of lnw and uerate against exper.
4. For the Personal Social Network data:
   a) Using Kruskal–Shepard metric MDS, lay out the social network in 3D, using the default settings. While MDS is running, play with the sliders for Data power and Weight, and note the changing impact of short and long distances. With MDS still running, activate GGobi's Move points Interaction mode, and experiment with grabbing points and moving them around the screen.
      i. Who is at the center of the network?
      ii. Name two other people who connect the center person with at least 5 other people.
      iii. Name the two people connected to Nikki but to nobody else.
      iv. Is there a relationship between maritalstat and position in the network?
      v. Does the central person work part time or full time?
   b) Choose a radial layout. Open a linked 1D ASH of log10(interactions), and find the pairs of people who have the greatest number of interactions. What do you find?
5. For the Music data:
   a) Use this code to get the dissimilarity matrix into GGobi, and set it up for MDS:
      ```
      > library(rggobi)
      > d.music <- read.csv("music-sub.csv", row.names=1)
      > f.std.data <- function(x){(x-mean(x))/sd(x)}
      > x <- apply(d.music[,3:7],2,f.std.data)
      ```

```
> d.music.cor <- cor(t(x))
> d.music.dist <- 1-abs(d.music.cor)
> gg <- ggobi(d.music)
> gg["distance"] <- as.vector(d.music.dist)
> names <- unlist(dimnames(d.music)[1])
> tmp <- matrix(c(rep(names,each=62),rep(names,62)),
    ncol=2)
> edges(gg["distance"]) <- tmp
```
  b) Run the Kruskal–Shepard metric MDS in 2D. Describe what you see. You may find it useful to brush the Rock, Classical, and New Wave tracks different colors or to add labels to the plot.
  c) How does randomizing the initial configuration affect the final configuration?
  d) Explore the effect of changing the scaling measure.
  e) Is a 3D configuration better than a 2D configuration?
6. Characterize the pattern of the labels in the $n = 4$ adjacent transposition graph. Lay out the $n = 5$ adjacent transposition graph and describe it.

7
Datasets

7.1 Tips

Source: Bryant, P. G. and Smith, M. A. (1995), *Practical Data Analysis: Case Studies in Business Statistics*, Richard D. Irwin Publishing, Homewood, IL.

Number of cases: 244
Number of variables: 8

Description: Food servers' tips in restaurants may be influenced by many factors, including the nature of the restaurant, size of the party, and table locations in the restaurant. Restaurant managers need to know which factors matter when they assign tables to food servers. For the sake of staff morale, they usually want to avoid either the substance or the appearance of unfair treatment of the servers, for whom tips (at least in restaurants in the United States) are a major component of pay.

In one restaurant, a food server recorded the following data on all customers they served during an interval of two and a half months in early 1990. The restaurant, located in a suburban shopping mall, was part of a national chain and served a varied menu. In observance of local law the restaurant offered seating in a non-smoking section to patrons who requested it. Each record includes a day and time, and taken together, they show the server's work schedule.

Variable	Explanation
obs	Observation number
totbill	Total bill (cost of the meal), including tax, in US dollars
tip	Tip (gratuity) in US dollars
sex	Sex of person paying for the meal (0=male, 1=female)
smoker	Smoker in party? (0=No, 1=Yes)
day	3=Thur, 4=Fri, 5=Sat, 6=Sun
time	0=Day, 1=Night
size	Size of the party

Primary question. What are the factors that affect tipping behavior?

Data restructuring: A new variable tiprate = tip/totbill should be calculated.

Analysis notes: This dataset is fabulously simple and yet fascinating. The original case study fits a traditional regression model, using tiprate as a response variable. The only important variable emerging from this model is size: As size increases, tiprate decreases. The reader may have noticed that restaurants seem to know about this association, because they often include a service charge for larger dining parties. (There has been at least one lawsuit regarding this service charge.) Here, this association explains only 2% of all the variation in tip rate — it is a very weak model! There are many other interesting features in the data, as described in this book.

Data files:
tips.csv, tips.xml

7.2 Australian Crabs

Source: Campbell, N. A. & Mahon, R. J. (1974), A Multivariate Study of Variation in Two Species of Rock Crab of genus *Leptograpsus*, *Australian Journal of Zoology* **22**, 417–425. The data was first brought to our attention by Venables & Ripley (2002) and Ripley (1996).

Number of rows: 200
Number of variables: 8

Description: Measurements on rock crabs of the genus *Leptograpsus*. One species *L. variegatus* had been split into two new species, previously grouped by color, orange and blue. Preserved specimens lose their color, so it was hoped that morphological differences would enable museum specimens to be classified. There are 50 specimens of each sex of each species, collected on site at Fremantle, Western Australia. For each specimen, five measurements were made, using vernier calipers.

Variable	Explanation
species	orange or blue
sex	male or female
index	1–200
frontal lip (FL)	length, in mm
rear width (RW)	width, in mm
carapace length (CL)	length of midline of the carapace, in mm
carapace width (CW)	maximum width of carapace, in mm
body depth (BD)	depth of the body; for females, measured after displacement of the abdomen, in mm

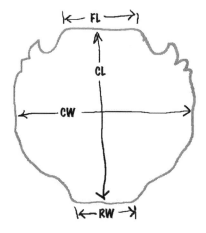

Primary question: Can we determine the species and sex of the crabs based on these five morphological measurements?

Data restructuring: A new class variable distinguishing all four groups would be useful.

Analysis notes: All physical measurements on the crabs are strongly positively correlated, and this is the main structure in the data. For this reason, it may be helpful to sphere the data and use principal components instead of raw variables in any analysis. Despite this strong association, there are a lot of differences among the four groups. Species can be perfectly distinguished by physical characteristics, and so can the sex of the larger crabs. In previous analyses, the measurements were logged, but we have not found this to be necessary.

Data files:
australian-crabs.csv, australian-crabs.xml

7.3 Italian Olive Oils

Source: Forina, M., Armanino, C., Lanteri, S. & Tiscornia, E. (1983), Classification of Olive Oils from their Fatty Acid Composition, in Martens, H. and Russwurm Jr., H., eds, Food Research and Data Analysis, Applied Science Publishers, London, pp. 189–214. It was brought to our attention by Glover & Hopke (1992).

Number of rows: 572
Number of variables: 10

Description: This data consists of the percentage composition of fatty acids found in the lipid fraction of Italian olive oils. The data arises from a study to determine the authenticity of an olive oil.

Variable	Explanation
region	Three "super-classes" of Italy: North, South, and the island of Sardinia
area	Nine collection areas: three from the region North (Umbria, East and West Liguria), four from South (North and South Apulia, Calabria, and Sicily), and two from the island of Sardinia (inland and coastal Sardinia).
palmitic, palmitoleic, stearic, oleic, linoleic, linolenic, arachidic, eicosenoic	fatty acids, % × 100

Primary question: How do we distinguish the oils from different regions and areas in Italy based on their combinations of the fatty acids?

Data restructuring: None needed.

Analysis notes: There are nine classes (areas) in this data, too many to easily classify. A better approach is to take advantage of the hierarchical structure in the data, partitioning by region before starting.

Some of the classes are easy to distinguish, but others present a challenge. The clusters corresponding to classes all have different shapes in the eight-dimensional data space.

Data files:
olive.csv, olive.xml

7.4 Flea Beetles

Source: Lubischew, A. A. (1962), On the Use of Discriminant Functions in Taxonomy, Biometrics **18**, 455–477.

Number of rows: 74
Number of variables: 7

Description: This data contains physical measurements on three species of flea beetles.

Variable	Explanation
species	*Ch. concinna, Ch. heptapotamica,* and *Ch. heikertingeri*
tars1	width of the first joint of the first tarsus in microns
tars2	width of the second joint of the first tarsus in microns
head	the maximal width of the head between the external edges of the eyes in 0.01 mm
aede1	the maximal width of the aedeagus in the fore-part in microns
aede2	the front angle of the aedeagus (1 unit = 7.5 degrees)
aede3	the aedeagus width from the side in microns

Primary question: How do we classify the three species?

Data restructuring: None needed.

Analysis notes: This straightforward dataset has three very well separated elliptically shaped clusters. It is fun to cluster the data with various algorithms and see how many get the clusters wrong.

Data files:
   flea.csv, flea.xml

7.5 PRIM7

Source: First used in Friedman, J. H. & Tukey, J. W. (1974), A Projection Pursuit Algorithm for Exploratory Data Analysis, *IEEE Transactions on Computing C* **23**, 881–889. Originally from Ballam, J., Chadwick, G. B., Guiragossian, G. T., Johnson, W. B., Leith, D. W. G. S., & Moriyasu, K. (1971), Van hove analysis of the reaction $\pi^- p \longrightarrow \pi^- \pi^- \pi^+ p$ and $\pi^+ p \longrightarrow \pi^+ \pi^+ \pi^- p$ at 16 gev/c*, *Physics Review D* **4**(1), 1946–1966.

Number of cases: 500
Number of variables: 7

Description: This data contains observations taken from a high-energy particle physics scattering experiment that yielded four particles. The reaction $\pi_b^+ p_t \to p\pi_1^+\pi_2^+\pi^-$ can be described completely by seven independent measurements. Below, $\mu^2(A, B, \pm C) = (E_A + E_B \pm E_C)^2 - (P_A + P_B \pm P_C)^2$ and $\mu^2(A, \pm B) = (E_A \pm E_B)^2 - (P_A \pm P_B)^2$, where E and P represent the particle's energy and momentum, respectively, as measured in billions of electron volts. The notation $(p)^2$ represents the inner product P/P. The ordinal assignment of the two π^+'s was done randomly.

Variable	Explanation
X1	$\mu^2(\pi^-, \pi_1^+, \pi_2^+)$
X2	$\mu^2(\pi^-, \pi_1^+)$
X3	$\mu^2(p, \pi^-)$
X4	$\mu^2(\pi^-, \pi_2^+)$
X5	$\mu^2(p, \pi_1^+)$
X6	$\mu^2(p, \pi_1^+, -p_t)$
X7	$\mu^2(p, \pi_2^+, -p_t)$

Primary question: What are the clusters in the data?

Data restructuring: None needed, although it is helpful to sphere the data to principal component coordinates before using projection pursuit.

Analysis notes: The case study illustrates the strength of our graphical methods in detecting sparse structure in high-dimensional space. It is a stunning look at uncovering a very geometric structure in high-dimensional space. A combination of interactive brush controls and motion graphics reveals that the points lie on a structure comprising connected low-dimensional pieces: a two-dimensional triangle, with two linear pieces extending from each vertex (Cook et al. 1995). Various graphical tools specifically facilitated the discovery of the structure: Plots of low-dimensional projections of the seven-dimensional object allowed discovery of the low-dimensional pieces, highlighting allowed the pieces to be recorded or marked, and animating many projections into a movie over time allowed the pieces to be reconstructed into the full shape. The data was 20 years old by the time these visual methods were applied to it, and the structure is known by physicists.

Data files:
    prim7.csv, prim7.xml

7.6 Tropical Atmosphere-Ocean Array (TAO)

Source: The data from the array, along with current updates, can be viewed on the web at http://www.pmel.noaa.gov/tao.

Number of cases: 736
Number of variables: 8

Description: The El Niño/Southern Oscillation (ENSO) cycle of 1982–1983, the strongest of the century, created many problems throughout the world. Parts of the world such as Peru and the United States experienced destructive flooding from increased rainfall, whereas countries in the western Pacific experienced drought and devastating brush fires. The ENSO cycle was neither predicted nor detected until it was near its peak, which highlighted the need for an ocean observing system to support studies of large-scale ocean-atmosphere interactions on seasonal-to-interannual time scales.

This observing system was developed by the international Tropical Ocean Global Atmosphere (TOGA) program. The Tropical Atmosphere Ocean (TAO) array consists of nearly 70 moored buoys spanning the equatorial Pacific, measuring oceanographic and surface meteorological variables critical for improved detection, understanding and prediction of seasonal-to-interannual climate variations originating in the tropics, most notably those related to the ENSO cycles.

The moorings were developed by the National Oceanic and Atmospheric Administration's (NOAA) Pacific Marine Environmental Laboratory (PMEL). Each mooring measures air temperature, relative humidity, surface winds, sea surface temperatures, and subsurface temperatures down to a depth of 500 meters, and a few of the buoys measure currents, rainfall, and solar radiation.

The TAO array provides real-time data to climate researchers, weather prediction centers, and scientists around the world. Forecasts for tropical Pacific Ocean temperatures for one to two years in advance can be made using the ENSO cycle data. These forecasts are possible because of the moored buoys, along with drifting buoys, volunteer ship temperature probes, and sea level measurements.

Variable	Explanation
year	1993 (a normal year), 1997 (an El Niño year), for November, December, and the January of the following year.
latitude	0°, 2°S, 5°S only.
longitude	110°W, 95°W only.
sea surface temp (SST)	measured in °C, at 1 m below the surface
air temp (AT)	measured in °C, at 3 m above the sea surface.
humidity (Hum)	relative humidity, measured 3 m above the sea surface.
uwind	east–west component of wind, measured 4 m above sea surface: positive means the wind is blowing toward the east.
vwind	north–south component of the wind, measured 4 m above sea surface: a positive sign means that the wind is blowing toward the north.

Primary question: Can we detect the El Niño event, based on sea surface temperature? What changes in the other observed variables occur during this event?

Data restructuring: This subset comes from a larger dataset extracted from the web site mentioned above. That data runs from March 7, 1980 to December 31, 1998, from −10°S to 10°N, and from 130°E to 90°W. There are 178,080 recorded measurements, with time, latitude, longitude, and five atmospheric variables for each record.

Longitude is measured with east as positive and west as negative units, with the prime meridian in Greenwich, UK, at 0°. The buoys are moored in the Pacific Ocean on the other side of the globe, with measurements on either side of the International Date Line (−180° = 180°)! This makes it very difficult to plot the data using numerical scales. We added new categorical variables marking the moored location of the buoys, making it easier to plot the spatial coordinates. These variables also make it easier to identify the buoys, because they tend to break free of the moorings and drift occasionally.

Analysis notes: One hurdle in working with this data is the large number of missing values. The missingness needs to be explored as a first step, and missing values need to be imputed before an analysis.

The larger data is cumbersome to work with, because of the missing values and the spatiotemporal context, but it has some interesting features. Plotting the latitude and longitude reveals that some buoys tend to drift, quite substantially at times, and that they are eventually retrieved and reattached to the moorings!

There is a massive El Niño event in the last year of this larger subset, 1997–1998, and it is visible at some locations when time series of sea surface temperature are plotted. Smaller El Niño events are visible at several other years. Changes in other variables are noticeable during these events too, particularly in one of the wind components.

Data files:

    tao.csv, tao.xml   Small subsets mostly used in the book.

    tao-full.csv,   The full data, not commonly used in the book, but
    tao-full.xml   included for data context.

7.7 Primary Biliary Cirrhosis (PBC)

Source: Distributed with Fleming & Harrington, *Counting Processes and Survival Analysis*, Wiley, New York, 1991, and available from http://lib.stat.cmu.edu/datasets. A description of the clinical background for the trial and the covariates recorded here is in Chapter 0, especially Section 0.2. It was originally from the Mayo Clinic trial in primary biliary cirrhosis (PBC) of the liver conducted between 1974 and 1984. A more extended discussion can be found in Dickson et al., Prognosis in primary biliary cirrhosis: model for decision making, *Hepatology* **10**, 1–7 (1989) and in Markus et al., Efficiency of liver transplantation in patients with primary biliary cirrhosis, *New England Journal of Medicine* **320**, 1709–1713 (1989).

Number of cases: 312
Number of variables: 20

Description: A total of 424 PBC patients, referred to the Mayo Clinic during that ten-year interval, met eligibility criteria for the randomized placebo controlled trial of the drug D-penicillamine. The first 312 cases in the dataset refer to subjects who participated in the randomized trial; they contain largely complete data. The additional 112 subjects did not participate in the clinical trial but consented to have basic measurements recorded and to be followed for survival; they are not represented here.

Variable	Explanation
id	
fu.days	number of days between registration and the earlier of death, transplantation, or study analysis time in July 1986
status	status is coded as 0=censored, 1=censored due to liver tx, 2=death
drug	1=D-penicillamine, 2=placebo
age	in days
sex	0=male, 1=female
ascites	presence of ascites: 0=no 1=yes
hepatom	presence of hepatomegaly: 0=no 1=yes
spiders	presence of spiders: 0=no 1=yes
edema	presence of edema: 0=no edema and no diuretic therapy for edema; .5 = edema present without diuretics, or edema resolved by diuretics; 1 = edema despite diuretic therapy
bili	serum bilirubin in mg/dl
chol	serum cholesterol in mg/dl
albumin	in gm/dl
copper	urine copper in μg/day
alk.phos	alkaline phosphatase in U/l
sgot	SGOT in U/ml
trig	triglycerides in mg/dl
platelet	platelets per cubic ml/1,000
protime	prothrombin time in seconds
stage	histologic stage of disease

Primary question: How do the different drugs affect the patients?

Data restructuring: Only records corresponding to patients that were in the original clinical trial were included in this data. The remaining records had too many systematic missing values.

Analysis notes: Handling missing values is an interesting exercise in this data, and experimenting with data transformations.

Data files:
 pbc.csv

7.8 Spam

Source: This was data collected at Iowa State University (ISU) by the 2003 Statistics 503 class.

7.8 Spam

Number of cases: 2,171
Number of variables: 21

Description: Every person monitored their email for a week and recorded information about each email message; for example, whether it was spam, and what day of the week and time of day the email arrived. We want to use this information to build a spam filter, a classifier that will catch spam with high probability but will never classify good email as spam.

Variable	Explanation
isuid	Iowa State U. student id (1–19)
id	email id (a unique message descriptor)
day of week	sun, mon, tue, wed, thu, fri, sat
time of day	0–23 (only integer values)
size.kb	size of email in kilobytes
box	yes if sender is in recipient's in- or outboxes (i.e., known to recipient); else no
domain	high-level domain of sender's email address: e.g., .edu, .ru
local	yes if sender's email is in local domain, else no; local addresses have the form xx@yy.iastate.edu
digits	number of numbers (0-9) in the sender's name: e.g., for lottery2003@yahoo.com, this is 4.
name	"name" (if first and last names are present), "single" (if only one name is present), or empty
capct	% capital letters in subject line
special	number of non-alphanumeric characters in subject
credit	yes if subject line includes one of mortgage, sale, approve, credit; else no
sucker	yes if subject line includes one of the words earn, free, save; else no
porn	yes if subject line includes one of nude, sex, enlarge, improve; else no
chain	yes if subject line includes one of pass, forward, help; else no
username	yes if subject includes recipient's name or login; else no
large.text	yes if email is HTML® and includes test for large font, defined as size = +3 or size = 5 or higher; else no
spampct	probability of being spam, according to ISU spam filter.
category	extended spam/mail category: "com," "list," "news," "ord"
spam	yes if spam; else no

Primary question: Can we distinguish between spam and "ham?"

Data restructuring: A *lot* of work was done to prepare this data for analysis! It is now quite clean, and no restructuring should be needed.

Analysis notes: The ISU mail handlers examine each email message and assign it a probability of being spam. Commonly used mail readers can use this information to file email directly into the trash, or at least to a special folder. It will be interesting to compare the results of a spam filter built on our collected data with results of the university's algorithm. (The university's algorithm was classifying a lot of email from the university president as spam for a short period!) Another aside is that there is a temporal trend to spam, which seems to be more frequent at some times of day and night. We have also seen that some users get more spam than others.

Careful choice of variables is needed for building the spam filter. Only those that might be automatically calculated by a mail handler are appropriate.

There are some missing values in the data due to differences between mail handlers and the availability of information about the emails.

Spammers evolve their attacks quickly, and the recognizable signs of spam of 2003 no longer exist in 2006 spam. For example, all spam now arrives with complete Caucasian-style name fields, and messages are embedded in images rather than plain text.

Data files:
spam.csv, spam.xml

7.9 Wages

Source: Singer, J. D. & Willett, J. B. (2003), *Applied Longitudinal Data Analysis*, Oxford University Press, Oxford, UK. It is a subset of data collected in the National Longitudinal Survey of Youth (NLSY) described at http://www.bls.gov/nls/nlsdata.htm.

Number of subjects: 888
Number of variables: 15
Number of observations, across all subjects: 6,402

Description: The data was collected to track the labor experiences of male high-school dropouts. The men were between 14 and 17 years old at the time of the first survey.

Variable	Explanation
id	1–888, for each subject.
lnw	natural log of wages, adjusted for inflation, to 1990 dollars.
exper	length of time in the workforce (in years). This is treated as the time variable, with t_0 for each subject starting on their first day at work. The number of time points and values of time points for each subject can differ.
ged	when/if a graduate equivalency diploma is obtained.
black	categorical indicator of race = black.
hispanic	categorical indicator of race = hispanic.
hgc	highest grade completed.
uerate	unemployment rates in the local geographic region at each measurement time.

Primary question: How do wages change with workforce experience?

Data restructuring: The data in its original form looked as follows, where time-independent variables have been repeated for each time point:

id	lnw	exper	black	hispanic	hgc	uerate
31	1.491	0.015	0	1	8	3.215
31	1.433	0.715	0	1	8	3.215
31	1.469	1.734	0	1	8	3.215
31	1.749	2.773	0	1	8	3.295
31	1.931	3.927	0	1	8	2.895
31	1.709	4.946	0	1	8	2.495
31	2.086	5.965	0	1	8	2.595
31	2.129	6.984	0	1	8	4.795
36	1.982	0.315	0	0	9	4.895
36	1.798	0.983	0	0	9	7.400
36	2.256	2.040	0	0	9	7.400
36	2.573	3.021	0	0	9	5.295
36	1.819	4.021	0	0	9	4.495
36	2.928	5.521	0	0	9	2.895
36	2.443	6.733	0	0	9	2.595
36	2.825	7.906	0	0	9	2.595
36	2.303	8.848	0	0	9	5.795
36	2.329	9.598	0	0	9	7.600

It was restructured into two tables of data. One table contains the time-independent measurements identified by subject id, and the other table contains the time-dependent variables:

id	black	hispanic	hgc
31	0	1	8
36	0	0	9

id	lnw	exper	uerate
31	1.491	0.015	3.215
31	1.469	1.734	3.215
31	1.749	2.773	3.295
31	1.931	3.927	2.895
31	1.709	4.946	2.495
31	2.086	5.965	2.595
31	2.129	6.984	4.795
36	1.982	0.315	4.895
36	1.798	0.983	7.400
36	2.256	2.040	7.400
36	2.573	3.021	5.295
36	1.819	4.021	4.495
36	2.928	5.521	2.895
36	2.443	6.733	2.595
36	2.825	7.906	2.595
36	2.303	8.848	5.795
36	2.329	9.598	7.600

Analysis notes: Singer & Willett (2003) use this data to illustrate fitting mixed linear models to ragged time indexed data. The analysis reports that the average growth in wages is about 4.7% for each year of experience. There is no difference between whites and Hispanics, but a big difference from blacks. The model uses a linear trend (on the log wages) to follow these patterns. The within-variance component of the model is significant, which indicates that the variability for each person is dramatically different. It does not tell us, however, in what ways people differ, and which people are similar.

The data is fascinating from a number of perspectives. Although on average wages tend to increase with time, the temporal patterns of individual wages vary dramatically. Some men experience a decline in wages over time, others a more satisfying increase, and yet others have very volatile wage histories. There is also a strange pattern differentiating the wage histories of black men from whites and Hispanics.

Data files:
    wages.xml

7.10 Rat Gene Expression

Source: X. Wen, S. Fuhrman, G. S. Michaels, D. B. Carr, S. Smith, J. L. Barker & R. Somogyi (1998), Large-scale temporal gene expression mapping of central nervous system development, in *Proceedings of the National Academy of Science* **95**, pp. 334–339, available on the web from http://pnas.org.

7.10 Rat Gene Expression

Number of cases: 112
Number of variables: 17

Description: This small subset of data is from a larger study of rat development using gene expression. The subset contains gene expression for nine developmental times, taken by averaging several replicates and normalizing the values using the maximum value for the gene.

Variable	Explanation
E11	11-day-old embryo
E13	13-day-old embryo
E15	15-day-old embryo
E18	18-day-old embryo
E21	21-day-old embryo
P0	at birth
P7	at 7 days
P14	at 14 days
A	adult
Class1 (*Class2*)	Functional classes representing expert's best guess: 1 neuro-glial markers *(1 markers)*, 2 neurotransmitter metabolizing enzymes *(2 neurotransmitter receptors, 3 GABA-A receptors, 4 glutamate receptors, 5 acetylcholine receptors, 6 serotonin receptors)*, 3 peptide signaling *(7 neurotrophins, 8 heparin-binding growth factors, 9 insulin/IGF)*, 4 diverse *(10 intracellular signaling, 11 cell cycle, 12 transcription factor, 13 novel/EST, 14 other)*
avcor	average linkage, correlation distance
wardcor	Wards linkage, correlation distance
comcor	complete linkage, correlation distance
avfluor	average linkage, fluorescence distance
wardfluor	Wards linkage, fluorescence distance
comfluor	complete linkage, fluorescence distance

Primary question: Do genes within a functional class have similar gene expression patterns? How does a clustering of genes compare with the functional classes?

Data restructuring: The data has been cleaned and heavily processed. The variables summarizing the cluster analysis were added, but beyond this no more restructuring of the data should be needed.

Analysis notes: The variables are time-ordered so parallel coordinate plots are very useful here. Brushing, particularly automatically from R, to focus on one

168   7 Datasets

functional class, or cluster, at a time is useful to compare patterns of gene expression between groups.

Data files:
ratsm.csv, ratsm.xml

7.11 Arabidopsis Gene Expression

Source: The data was collected in Dr. Basil Nikolau's lab at Iowa State University and it is discussed in Cook, Hofmann, Lee, Yang, Nikolau & Wurtele (2007).

Number of cases: 8,297
Number of variables: 9

Description: This data is from a two-factor, single replicate experiment of the following form:

	Treatment added	
	no	yes
Mutant	*M1,M2*	MT1, MT2
Wild type	W1, W2	WT1, WT2

The mutant organism is defective in the ability to synthesize an essential cofactor, which is provided by the treatment.

The data was recorded on Affymetrix GeneChip Arabidopsis Genome Arrays. The raw data was processed using robust median average and quantiles normalization available in the Bioconductor suite of tools (Bioconductor 2006).

Variable	Explanation
Gene ID	Affymetrix unique identifier for each gene. This is used as a label in the data, and for linking between multiple forms of the data.
M1	Mutant, no treatment added, replicate 1
M2	Mutant, no treatment added, replicate 2
MT1	Mutant, treatment added, replicate 1
MT2	Mutant, treatment added, replicate 2
WT1	Wild type, no treatment added, replicate 1
WT2	Wild type, no treatment added, replicate 2
WTT1	Wild type, treatment added, replicate 1
WTT2	Wild type, treatment added, replicate 2

7.11 Arabidopsis Gene Expression

Primary question: Which genes are differentially expressed when the treatment is not added, with special interest in the mutant genotype?

Data restructuring: Two forms are provided in different tables of data so that we can examine the replicate data values in association with the overall variation.

GeneID	M1 M2	MT1 MT2	WT1 WT2	WTT1 WTT2
1				
⋮				
8297				

GeneID	M	MT	WT	WTT
1				
⋮				
8297				
1				
⋮				
8297				

Averages across replicates are added to the short form of the data.

Summaries from ANOVA models fit for each gene in the data are included in the short form. These are useful for helping to detect differentially expressed genes.

Analysis notes: Difference is measured by how the individual gene varies in the replicate and by how all the genes vary in expression value.

We would also hope to see (1) small differences in expression values in the replications, (2) small differences between expression values in wild type with and without the treatment added, and (3) little difference between expression values in the mutant with the treatment and wild type.

It is important to emphasize the difference in analysis of microarray data from many other multivariate data analysis tasks. In microarray data, it is important to find a small number of genes that are behaving differently from others in an understandable way. This task involves both multiple comparisons and outlier detection. From the perspective of a traditional statistical analysis, we are merely dealing with a problem that could be solved by a t-test for comparing means. The drawback is that we have to do a test for every single gene!

Conventionally this type of data is plotted using a heatmap, shown below, but a lot of information can be obtained from linked scatterplots and parallel coordinate plots.

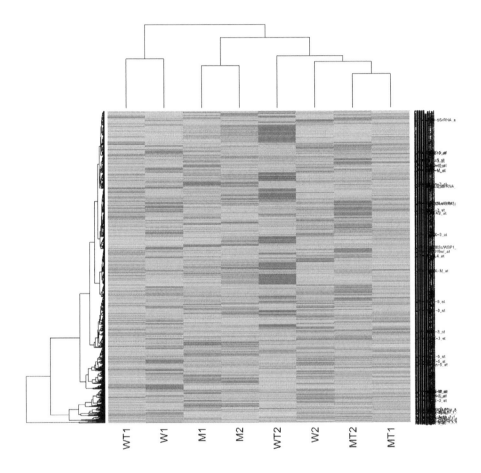

This field of research is evolving rapidly, and data and analysis methods change frequently.

Data files:

arabidopsis.xml

7.12 Music

Source: Collected by Dianne Cook.

Number of cases: 62
Number of variables: 7

Description: Using an Apple computer, each track was read into the music editing software Amadeus II, and the first 40-second clip was snipped and saved as a WAV file. (WAV is an audio format developed by Microsoft®, commonly used on Windows but becoming less popular.) These files were read into R using the package tuneR (Ligges 2006), which converts the audio file into numeric data. All of the CDs contained left and right channels, and variables were calculated on both channels.

Variable	Explanation
artist	Abba, Beatles, Eels, Vivaldi, Mozart, Beethoven, Enya
type	rock, classical, or new wave
lvar, lave, lmax	average, variance, maximum of the frequencies of the left channel
lfener	an indicator of the amplitude or loudness of the sound
lfreq	median of the location of the 15 highest peak in the periodogram

Primary question: Can we distinguish between rock and classical tracks? Can we group the tracks into a small number of clusters according to their similarity on audio characteristics?

Data restructuring: This dataset is very clean and simplified. The original data contained 72 variables, most of which have been excluded.

Analysis notes: Answers to the primary question might be used to arrange tracks on a digital music player or to make recommendations. Other questions of interest might be:

- Do the rock tracks have different characteristics than classical tracks?
- How does Enya compare with rock and classical tracks?
- Are there differences between the tracks of different artists?

Data files:

`music-sub.csv,` `music-sub.xml`	Subset of data used in this book. The last five tracks in the data (58–62) have the artist and type of music loosely disguised so that they can be used to test classifiers that students built using the rest of the data.
`music-all.csv,` `music-all.xml`	Full datasets, 72 variables, and a few missing values.
`music-clust.csv,` `music-clust.xml`	Subset of data, augmented with results from different cluster analyses
`music-SOM1.xml,` `music-SOM2.xml`	Different SOM models appended to the data.

7.13 Cluster Challenge

Source: Simulated by Dianne Cook.

Number of cases: 250
Number of variables: 5
Description: Simulated data included as a challenge to find the number of clusters.

Primary question: How many clusters in this data?

Data files:
   `clusters-unknown.csv`

7.14 Adjacent Transposition Graph

Source: Constructed by Deborah F. Swayne.

Number of cases: 24 nodes and 36 edges in the $n = 4$ adjacent transposition graph; 120 nodes and 240 edges in the $n = 5$ graph.
Number of variables: 3 variables in the $n = 4$ graph; 4 variables in the $n = 5$ graph.
Description: The $n = N$ adjacent transposition graph is generated as follows. Start with all permutations of the sequence 1, 2, ..., N. There are $N!$ such sequences; make each one a vertex in the graph. Connect two vertices by

an edge if one permutation can be turned into the other by transposing two adjacent elements.

Principal question: Can a graph layout algorithm be used to arrange the nodes so that it is easy to understand the different permutations of rankings?

Data files:
   adjtrans4.xml, adjtrans5.xml

7.15 Florentine Families

Source:

This data is widely known within the social network community, and is readily available from a number of sources. It was compiled by John Padgett from historical documents such as Kent (1978). The 16 families were chosen for analysis from a much larger collection of 116 leading Florentine families because of their historical prominence. Padgett and Ansell (1993) and Breiger and Pattison (2006) extensively analyzed the data.

We obtained it from the R package SNAData, by D. Scholtens (2006), part of the Bioconductor project; Scholtens obtained it from Wasserman & Faust (1994).

Number of cases: 16 nodes; two sets of edges, one with 15 edges and the other with 20.

Number of variables: 3 variables on each node; one on each edge.

Description:

The data include families who were locked in a struggle for political control of the city of Florence in around 1430. Two factions were dominant in this struggle: one revolved around the infamous Medicis, and the other around the powerful Strozzis.

Variable	Explanation
Wealth	Family net wealth in 1427 (in thousands of lira)
NumberPriorates	Number of seats on the civic council held by the family between 1282 and 1344
NumberTies	Total number of business or marriage ties
AveNTies in Business and Marital tables	Average number of business (loans, credits, joint partnerships) or marital ties per family

Primary question: How are the dominant families of old Florence connected to each other?

Data files:
   FlorentineFam.xml, constructed from SNAData.

7.16 Morse Code Confusion Rates

Source: Rothkopf, E. Z. (1957), A Measure of Stimulus Similarity and Error in some Paired-Associate Learning Tasks, *Journal of Experimental Psychology* **53**, 94–101.

Number of pairwise distances: 1,260

Description:

In an experiment, inexperienced subjects were exposed to pairs of Morse codes in rapid order. The subjects had to decide whether the two codes in a pair were identical. The data were summarized in a table of confusion rates.

Confusion rates are similarity measures: Codes that are often confused are interpreted as "similar" or "close." Similarity measures are converted to dissimilarity measures so that multidimensional scaling can be applied.

Morse codes consist of sequences of short and long sounds, which are called "dots" and "dashes" and written using the characters "." and "-". Examples are:

Letter	Code	Letter	Code	Digit	Code
A	. -	F	. . - .	1	. - - - -
B	- . . .	G	- - .	2	. . - - -
C	- . - .	H	. . . .		
D	- . .	T	-		
E	.	X	- . . -		

The codes are of varying length, with the shorter codes representing letters that are more common in English. The digits are all five-character codes.

Variable	Explanation
Length	Length of the code, rescaled to [0,1]
Dashes	Number of dashes
D	Dissimilarity between codes

Primary question: Which codes are similar and often confused with each other?

Data restructuring:

The original data came as an asymmetric 36×36 matrix of similarities, $S_{i,j}, i, j = 1, ..., n$. The values were converted to dissimilarities $D_{i,j}$ and symmetrized, using $D_{i,j}^2 = S_{i,i} + S_{j,j} - 2S_{i,j}$. Two variables were derived from the Morse codes themselves, **Length** and **Dashes**.

The dissimilarity matrix was reconfigured to conform to GGobi's XML format, a set of $n(n-1) = 1,260$ edges with associated dissimilarity. A second set of 33 edges was added to link similar codes; it is for display only, to aid in interpretation of the configuration, and is not used by the MDS algorithm.

Analysis notes: Start with the edges turned off, and focus on the movement of the points. Add the edges when the layout is complete to understand the structure of the final configuration. Re-start MDS a few times from random starting positions, and compare the resulting configurations.

Data files:
morsecodes.xml

7.17 Personal Social Network

Source: Provided by Chris Volinsky and Deborah F. Swayne.

Number of cases: 140 nodes (people) and 203 edges (contacts between people).
Number of variables: two categorical variables for each node; one categorical and two real variables for each edge.

Description: This is a *personal social network*, collected by selecting one person, adding that person's contacts, each contact's contacts, and so on. In its original form, the nodes were telephone numbers and the edges represented calls from one number to another (Cortes, Pregibon & Volinsky 2003), but the privacy of individuals has been protected by disguising the telephone numbers as names and changing the meaning of the original variables.

People:

Variable	Explanation
maritalstat	categorical: married, never married, or other
hours	binary: full time or part time

Contacts:

Variable	Explanation
interactions	a measure of the amount of time spent talking
center triangle	binary; is the point part of a 3-node cycle?
log10(interactions)	base 10 log of interactions

Data files:
snetwork.xml

References

Ahlberg, C., Williamson, C. & Shneiderman, B. (1991), Dynamic Queries for Information Exploration: An Implementation and Evaluation, in *ACM CHI '92 Conference Proceedings*, Association of Computing Machinery, New York, pp. 619–626. See also http://www.cs.umd.edu/hcil/pubs/tech-reports.shtml.

Andrews, D. F. (1972), "Plots of High-dimensional Data, *Biometrics* **28**, 125–136.

Anselin, L. & Bao, S. (1997), Exploratory Spatial Data Analysis Linking SpaceStat and ArcView, in M. M. Fischer & A. Getis, eds, *Recent Developments in Spatial Analysis*, Springer, Berlin, pp. 35–59.

Asimov, D. (1985), The Grand Tour: A Tool for Viewing Multidimensional Data, *SIAM Journal of Scientific and Statistical Computing* **6**(1), 128–143.

Ballam, J., Chadwick, G. B., Guiragossian, G. T., Johnson, W. B., Leith, D. W. G. S. & Moriyasu, K. (1971), Van hove analysis of the reaction $\pi^- p \longrightarrow \pi^- \pi^- \pi^+ p$ and $\pi^+ p \longrightarrow \pi^+ \pi^+ \pi^- p$ at 16 gev/c*, *Physics Review D* **4**(1), 1946–1966.

Becker, R. A. & Chambers, J. M. (1984), *S: An Environment for Data Analysis and Graphics*, Wadsworth, Belmont, CA.

Becker, R. A. & Cleveland, W. S. (1988), Brushing Scatterplots, in W. S. Cleveland & M. E. McGill, eds, *Dynamic Graphics for Statistics*, Wadsworth, Monterey, CA, pp. 201–224.

Becker, R., Cleveland, W. S. & Shyu, M.-J. (1996), The Visual Design and Control of Trellis Displays, *Journal of Computational and Graphical Statistics* **6**(1), 123–155.

Bederson, B. B. & Schneiderman, B. (2003), *The Craft of Information Visualization: Readings and Reflections*, Morgan Kaufmann, San Diego, CA.

Bioconductor (2006), Open Source Software for Bioinformatics, http://www.bioconductor.org.

Bishop, C. M. (2006), *Pattern Recognition and Machine Learning*, Springer, New York.

Bonneau, G., Ertl, T. & Nielson, G. M., eds (2006), *Scientific Visualization: The Visual Extraction of Knowledge from Data*, Springer, New York.

Borg, I. & Groenen, P. J. F. (2005), *Modern Multidimensional Scaling*, Springer, New York.

Boser, B. E., Guyon, I. M. & Vapnik, V. N. (1992), A Training Algorithm for Optimal Margin Classifiers, in COLT '92: Proceedings of the Fifth Annual Workshop on Computational Learning Theory, ACM Press, New York, pp. 144–152.

Breiger, R. & Pattison, P. (1986), Cumulated Social Roles. The Duality of Persons and their Algebras, *Social Networks* **8**, 215–256.

Breiman, L. (2001), Random Forests, *Machine Learning* **45**(1), 5–32.

Breiman, L. & Cutler, A. (2004), Random Forests, http://www.math.usu.edu/~adele/forests/cc_home.htm.

Breiman, L., Friedman, J., Olshen, C. & Stone, C. (1984), *Classification and Regression Trees*, Wadsworth and Brooks/Cole, Monterey, CA.

Bryant, P. G. & Smith, M. A. (1995), *Practical Data Analysis: Case Studies in Business Statistics*, Richard D. Irwin Publishing, Homewood, IL.

Buja, A. (1996), Interactive Graphical Methods in the Analysis of Customer Panel Data: Comment, *Journal of Business & Economic Statistics* **14**(1), 128–129.

Buja, A. & Asimov, D. (1986), Grand Tour Methods: An Outline, *Computing Science and Statistics* **17**, 63–67.

Buja, A. & Swayne, D. F. (2002), Visualization Methodology for Multidimensional Scaling, *Journal of Classification* **19**(1), 7–43.

Buja, A. & Tukey, P., eds (1991), *Computing and Graphics in Statistics*, Springer-Verlag, New York.

Buja, A., Cook, D. & Swayne, D. (1996), Interactive High-Dimensional Data Visualization, *Journal of Computational and Graphical Statistics* **5**(1), 78–99. See also www.research.att.com/~andreas/xgobi/heidel/.

Buja, A., Cook, D., Asimov, D. & Hurley, C. (2005), Computational Methods for High-Dimensional Rotations in Data Visualization, *in* C. R. Rao, E. J. Wegman & J. L. Solka, eds, *Handbook of Statistics: Data Mining and Visualization*, Elsevier/North-Holland, Amsterdam, The Netherlands, pp. 391–414.

Buja, A., Hurley, C. & McDonald, J. A. (1986), A Data Viewer for Multivariate Data, *Computing Science and Statistics* **17**(1), 171–174.

Buja, A., McDonald, J. A., Michalak, J. & Stuetzle, W. (1991), Interactive Data Visualization using Focusing and Linking, *in* G. M. Nielson & L. Rosenblum, eds, Proceedings of Visualization '91, IEEE Computer Society Press, Los Alamitos, CA, pp. 156–162.

Buja, A., Swayne, D. F., Littman, M. L., Dean, N. & Hofmann, H. (2007), Interactive Data Visualization with Multidimensional Scaling, *Journal of Computational and Graphical Statistics (under review)*. See http://www-stat.wharton.upenn.edu/~buja/PAPERS/paper-mds-jcgs.pdf.

Campbell, N. A. & Mahon, R. J. (1974), A Multivariate Study of Variation in Two Species of Rock Crab of Genus *leptograpsus*, *Australian Journal of Zoology* **22**, 417–425.

Card, S. K., Mackinlay, J. D. & Schneiderman, B. (1999), *Readings in Information Visualization*, Morgan Kaufmann Publishers, San Francisco, CA.

Carr, D. B., Wegman, E. J. & Luo, Q. (1996), ExplorN: Design Considerations Past and Present, Technical Report 129, Center for Computational Statistics, George Mason University, Fairfax, VA.

Chang, C.-C. & Lin, C.-J. (2006), LIBSVM: A Library for Support Vector Machines, http://www.csie.ntu.edu.tw/~cjlin/libsvm.

Chatfield, C. (1995), *Problem Solving: A Statistician's Guide*, Chapman and Hall/CRC Press, London.

Chen, C.-H., Härdle, W. & Unwin, A., eds (2007), *Handbook of Data Visualization*, Springer, Berlin.

Cheng, B. & Titterington, M. (1994), Neural Networks: A Review from a Statistical Perspective, *Statistical Science* **9**(1), 2–30.

Cleveland, W. S. (1993), *Visualizing Data*, Hobart Press, Summit, NJ.
Cleveland, W. S. & McGill, M. E., eds (1988), *Dynamic Graphics for Statistics*, Wadsworth, Monterey, CA.
Cook, D., Buja, A. & Cabrera, J. (1993), Projection Pursuit Indexes Based on Orthonormal Function Expansions, *Journal of Computational and Graphical Statistics* **2**(3), 225–250.
Cook, D., Buja, A., Cabrera, J. & Hurley, C. (1995), Grand Tour and Projection Pursuit, *Journal of Computational and Graphical Statistics* **4**(3), 155–172.
Cook, D., Hofmann, H., Lee, E.-K., Yang, H., Nikolau, B. & Wurtele, E. (2007), Exploring Gene Expression Data, Using Plots, *Journal of Data Science* **5**(2), 151–182.
Cook, D., Lee, E.-K., Buja, A. & Wickham, H. (2006), Grand Tours, Projection Pursuit Guided Tours and Manual Controls, *in* C.-H. Chen, W. Härdle & A. Unwin, eds, *Handbook of Data Visualization*, Springer, Berlin, Germany.
Cook, D., Majure, J. J., Symanzik, J. & Cressie, N. (1996), Dynamic Graphics in a GIS: Exploring and Analyzing Multivariate Spatial Data using Linked Software, *Computational Statistics: Special Issue on Computer Aided Analyses of Spatial Data* **11**(4), 467–480.
Cortes, C. & Vapnik, V. N. (1995), Support-Vector Networks, *Machine Learning* **20**(3), 273–297.
Cortes, C., Pregibon, D. & Volinsky, C. (2003), Computational Methods for Dynamic Graphs, *Journal of Computational & Graphical Statistics* **12**(4), 950–970.
Crowder, M. J. & Hand, D. J. (1990), *Analysis of Repeated Measures*, Chapman and Hall, London.
Dalgaard, P. (2002), *Introductory Statistics with R*, Springer, New York.
Dasu, T., Swayne, D. F. & Poole, D. (2005), Grouping Multivariate Time Series: A Case Study, *in* Proceedings of the IEEE Workshop on Temporal Data Mining: Algorithms, Theory and Applications, in conjunction with the Conference on Data Mining, Houston, November 27, 2005, IEEE Computer Society, pp. 25–32. See also http://www.cs.fiu.edu/ taoli/workshop/TDM2005/TDM-proceedings.pdf.
Di Battista, G., Eades, P., Tamassia, R. & Tollis, I. G. (1998), *Graph Drawing: Algorithms for the Visualization of Graphs*, Prentice Hall, Englewood Cliffs, NJ.
Dickson, E. R., Grambsch, P. M., Fleming, T. R., Fisher, L. D. & A, A. L. (1989), Prognosis in primary biliary cirrhosis: model for decision making, *Hepatology* **10**, 1–7.
Diggle, P. J., Heagerty, P. J., Liang, K.-Y. & Zeger, S. L. (2002), *Analysis of Longitudinal Data*, Oxford University Press, Oxford, UK.
Dimitriadou, E., Hornik, K., Leisch, F., Meyer, D. & Weingessel, A. (2006), e1071: Misc Functions of the Department of Statistics, TU Wien, http://www.R-project.org.
d'Ocagne, M. (1885), *Coordonnées Parallèles et Axiales: Méthode de transformation géométrique et procédé nouveau de calcul graphique déduits de la considération des coordonnées parallèles*, Gauthier-Villars, Paris, France.
Dupont, C. & Harrell, F. E. (2006), Hmisc: Harrell Miscellaneous, http://www.R-project.org.
Dykes, J., MacEachren, A. M. & Kraak, M.-J. (2005), *Exploring Geovisualization*, Elsevier, New York.

References

Everitt, B. S., Landau, S. & Leese, M. (2001), *Cluster Analysis (4th ed)*, Edward Arnold, London.

Fisher, R. A. (1936), The Use of Multiple Measurements in Taxonomic Problems, *Annals of Eugenics* **7**, 179–188.

Fleming, T. R. & Harrington, D. P. (1991), *Counting Processes and Survival Analysis*, John Wiley & Sons, New York.

Ford, B. J. (1992), *Images of Science: A History of Scientific Illustration*, The British Library, London.

Forina, M., Armanino, C., Lanteri, S. & Tiscornia, E. (1983), Classification of Olive Oils from their Fatty Acid Composition, *in* H. Martens & H. Russwurm Jr., eds, *Food Research and Data Analysis*, Applied Science Publishers, London, pp. 189–214.

Fraley, C. & Raftery, A. E. (2002), Model-based Clustering, Discriminant Analysis, Density Estimation, *Journal of the American Statistical Association* **97**, 611–631.

Friedman, J. H. & Tukey, J. W. (1974), A Projection Pursuit Algorithm for Exploratory Data Analysis, *IEEE Transactions on Computing C* **23**, 881–889.

Friendly, M. & Denis, D. J. (2006), Graphical Milestones, http://www.math.yorku.ca/SCS/Gallery/milestone.

Fua, Y.-H., Ward, M. O. & Rundensteiner, E. A. (1999), Hierarchical Parallel Coordinates for Exploration of Large Datasets, *in* VIS '99: Proceedings of the conference on Visualization '99, IEEE Computer Society Press, Los Alamitos, CA, USA, pp. 43–50.

Furnas, G. W. & Buja, A. (1994), Prosection Views: Dimensional Inference Through Sections and Projections, *Journal of Computational and Graphical Statistics* **3**(4), 323–385.

Gabriel, K. R. (1971), The Biplot Graphical Display of Matrices with Applications to Principal Component Analysis, *Biometrika* **58**, 453–467.

Gansner, E. R. & North, S. C. (2000), An Open Graph Visualization System and its Applications to Software Engineering, *Software — Practice and Experience* **30**(11), 1203–1233.

Glover, D. M. & Hopke, P. K. (1992), Exploration of Multivariate Chemical Data by Projection Pursuit, *Chemometrics and Intelligent Laboratory Systems* **16**, 45–59.

Good, P. (2005), *Permutation, Parametric, and Bootstrap Tests of Hypotheses*, Springer, New York.

Gower, J. C. & Hand, D. J. (1996), *Biplots*, Chapman and Hall, London.

Hansen, C. & Johnson, C. R. (2004), *Visualization Handbook*, Academic Press, Orlando, FL.

Hartigan, J. A. & Kleiner, B. (1981), Mosaics for Contingency Tables, *in* Computer Science and Statistics: Proceedings of the 13th Symposium on the Interface, Interface Foundation of North America, Inc., Fairfax Station, VA, pp. 268–273.

Hastie, T., Tibshirani, R. & Friedman, J. (2001), *The Elements of Statistical Learning*, Springer, New York.

Hofmann, H. (2001), *Graphical Tools for the Exploration of Multivariate Categorical Data*, Books on Demand, http://www.bod.de.

Hofmann, H. (2003), Constructing and Reading Mosaicplots, *Computational Statistics and Data Analysis* **43**(4), 565–580.

Hofmann, H. & Theus, M. (1998), Selection Sequences in MANET, *Computational Statistics* **13**(1), 77–87.

Hurley, C. (1987), The Data Viewer: An Interactive Program for Data Analysis, PhD thesis, University of Washington, Seattle.

Ihaka, R. & Gentleman, R. (1996), R: A Language for Data Analysis and Graphics, *Journal of Computational and Graphical Statistics* **5**, 299–314.

Inselberg, A. (1985), The Plane with Parallel Coordinates, *The Visual Computer* **1**, 69–91.

Joachims, T. (1999), Making Large-Scale SVM Learning Practical, *in* B. Schölkopf, C. Burges & A. Smola, eds, *Advances in Kernel Methods-Support Vector Learning*, MIT Press, Cambridge, MA.

Johnson, R. A. & Wichern, D. W. (2002), *Applied Multivariate Statistical Analysis (5th ed)*, Prentice-Hall, Englewood Cliffs, NJ.

Kent, D. (1978), *The Rise of the Medici: Faction in Florence, 1426-1434*, Oxford University Press, Oxford, UK.

Kohonen, T. (2001), *Self-Organizing Maps (3rd ed)*, Springer, Berlin.

Koschat, M. A. & Swayne, D. F. (1996), Interactive Graphical Methods in the Analysis of Customer Panel Data (with discussion), *Journal of Business and Economic Statistics* **14**(1), 113–132.

Kruskal, J. B. (1964a), Multidimensional Scaling by Optimizing Goodness of Fit to a Nonmetric Hypothesis, *Psychometrika* **29**, 1–27.

Kruskal, J. B. (1964b), Nonmetric Multidimensional Scaling: a Numerical Method, *Psychometrika* **29**, 115–129.

Kruskal, J. B. & Wish, M. (1978), *Multidimensional Scaling*, Sage Publications, London.

Lee, E. K., Cook, D., Klinke, S. & Lumley, T. (2005), Projection Pursuit for Exploratory Supervised Classification, *Journal of Computational and Graphical Statistics* **14**(4), 831–846.

Liaw, A., Wiener, M., Breiman, L. & Cutler, A. (2006), randomForest: Breiman and Cutler's Random Forests for Classification and Regression, http://www.R-project.org.

Ligges, U. (2006), tuneR: Analysis of Music, http://www.R-project.org.

Little, R. J. A. & Rubin, D. B. (1987), *Statistical Analysis with Missing Data*, John Wiley & Sons, New York.

Littman, M. L., Swayne, D. F., Dean, N. & Buja, A. (1992), Visualizing the Embedding of Objects in Euclidean Space, *in* Computing Science and Statistics: Proceedings of the 24th Symposium on the Interface, Interface Foundation of North America, Inc., Fairfax Station, VA, pp. 208–217.

Longley, P. A., Maguire, D. J., Goodchild, M. F. & Rhind, D. W. (2005), *Geographic Information Systems and Science*, John Wiley & Sons, New York.

Lubischew, A. A. (1962), On the Use of Discriminant Functions in Taxonomy, *Biometrics* **18**, 455–477.

Maindonald, J. & Braun, J. (2003), *Data Analysis and Graphics using R - An Example-based Approach*, Cambridge University Press, Cambridge.

Markus, B., Dickson, E., Grambsch, P., Fleming, T., Mazzaferro, V., Klintmalmd, G., Wiesner, R., Thiel, D. V. & Starzl, T. (1989), Efficiency of liver transplantation in patients with primary biliary cirrhosis, *New England Journal of Medicine* **320**, 1709–1713.

McFarlane, M. & Young, F. W. (1994), Graphical Sensitivity Analysis for Multidimensional Scaling, *Journal of Computational and Graphical Statistics* **3**, 23–33.

McNeil, D. (1977), *Interactive Data Analysis*, John Wiley & Sons, New York.

Murrell, P. (2005), *R Graphics*, Chapman & Hall/CRC, Boca Raton, FL.

Nie, H. H., Jenkins, J. G., Steinbrenner, K. & Bent, D. H. (1975), SPSS: Statistical Package for the Social Sciences, SPSS, Inc., Chicago, IL.

Novo, A. A. & Schafer, J. (2006), norm: Analysis of Multivariate Normal Datasets with Missing Values, http://www.R-project.org.

Padgett, J. F. & Ansell, C. K. (1993), Robust Action and the Rise of the Medici, 1400-1434, *American Journal of Sociology* **98**, 1259–1319.

R Development Core Team (2006), *R: A Language and Environment for Statistical Computing*, R Foundation for Statistical Computing, Vienna, Austria.

Rao, C. R. (1948), The Utilization of Multiple Measurements in Problems of Biological Classification (with discussion), *Journal of the Royal Statistical Society, Series B* **10**, 159–203.

Rao, C. R., ed. (1993), *Handbook of Statistics, Vol. 9*, Elsevier Science Publishers, Amsterdam, The Netherlands.

Ripley, B. (1996), *Pattern Recognition and Neural Networks*, Cambridge University Press, Cambridge.

Rothkopf, E. Z. (1957), A Measure of Stimulus Similarity and Errors in Some Paired-associate Learning Tasks, *Journal of Experimental Psychology* **53**, 94–101.

Schafer, J. L. (1997), *Analysis of Incomplete Multivariate Data*, Chapman and Hall, London.

Scholtens, D. (2006), SNAData, http://www.bioconductor.org.

Scott, D. W. (1992), *Multivariate Density Estimation: Theory, Practice, and Visualization*, John Wiley & Sons, New York.

Shepard, R. N. (1962), The Analysis of Proximities: Multidimensional Scaling with an Unknown Distance Function, I and II, *Psychometrika* **27**, 125–139 and 219–246.

Singer, J. D. & Willett, J. B. (2003), *Applied Longitudinal Data Analysis*, Oxford University Press, Oxford, UK.

Spence, R. (2007), *Information Visualization: Design for Interaction*, Prentice Hall.

Swayne, D. & Buja, A. (1998), Missing Data in Interactive High-Dimensional Data Visualization, *Computational Statistics* **13**(1), 15–26.

Swayne, D. F., Cook, D. & Buja, A. (1992), XGobi: Interactive Dynamic Graphics in the X Window System with a Link to S, *in* American Statistical Association 1991 Proceedings of the Section on Statistical Graphics, American Statistical Association, Alexandria, VA, pp. 1–8.

Swayne, D. F., Temple Lang, D., Buja, A. & Cook, D. (2003), GGobi: Evolving from XGobi into an Extensible Framework for Interactive Data Visualization, *Computational Statistics & Data Analysis* **43**, 423–444.

Takatsuka, M. & Gahegan, M. (2002), GeoVISTA Studio: A Codeless Visual Programming Environment for Geoscientific Data Analysis and Visualization, *The Journal of Computers and Geosciences* **28**(10), 1131–1144.

Temple Lang, D., Swayne, D., Wickham, H. & Lawrence, M. (2006), rggobi: An Interface between R and GGobi, http://www.R-project.org.

Theus, M. (2002), Interactive Data Visualization Using Mondrian, *Journal of Statistical Software* **7**(11), http://www.jstatsoft.org.

Thompson, G. L. (1993), Generalized Permutation Polytopes and Exploratory Graphical Methods for Ranked Data, *The Annals of Statistics* **21**, 1401–1430.

Tierney, L. (1991), *LispStat: An Object-Orientated Environment for Statistical Computing and Dynamic Graphics*, John Wiley & Sons, New York.

Torgerson, W. S. (1952), Multidimensional Scaling. 1. Theory and Method, *Psychometrika* **17**, 401–419.
Tufte, E. (1983), *The Visual Display of Quantitative Information*, Graphics Press, Cheshire, CT.
Tufte, E. (1990), *Envisioning Information*, Graphics Press, Cheshire, CT.
Tukey, J. W. (1965), The Technical Tools of Statistics, *The American Statistician* **19**, 23–28.
Tukey, J. W. & Tukey, P. (1990), Strips Displaying Empirical Distributions: I. Textured Dot Strips, Bellcore Technical Memorandum.
Unwin, A. R., Hawkins, G., Hofmann, H. & Siegl, B. (1996), Interactive Graphics for Data Sets with Missing Values - MANET, *Journal of Computational and Graphical Statistics* **5**(2), 113–122.
Unwin, A., Theus, M. & Hofmann, H. (2006), *Graphics of Large Datasets: Visualizing a Million*, Springer, Berlin.
Unwin, A., Volinsky, C. & Winkler, S. (2003), Parallel Coordinates for Exploratory Modelling Analysis, *Comput. Stat. Data Anal.* **43**(4), 553–564.
Urbanek, S. & Theus, M. (2003), iPlots: High Interaction Graphics for R, *in* K. Hornik, F. Leisch & A. Zeileis, eds, Proceedings of the 3rd International Workshop on Distributed Statistical Computing (DSC 2003), http://www.ci.tuwien.ac.at/Conferences/DSC-2003.
Vapnik, V. N. (1999), *The Nature of Statistical Learning Theory*, Springer, New York.
Velleman, P. F. & Velleman, A. Y. (1985), *Data Desk Handbook*, Data Description, Inc, Ithaca, NY.
Venables, W. N. & Ripley, B. (2002), *Modern Applied Statistics with S*, Springer-Verlag, New York.
Vlachos, P. (2006), StatLib: Data, Software and News from the Statistics Community, http://lib.stat.cmu.edu/datasets/.
Wainer, H. (2000), *Visual Revelations (2nd ed)*, LEA, Inc, Hillsdale, NJ.
Wainer, H. & Spence, I. e. (2005a), *The Commercial and Political Atlas, Representing, by means of Stained Copper-Plate Charts, The Progress of the Commerce, Revenues, Expenditure, and Debts of England, during the whole of the Eighteenth Century, by William Playfair*, Cambridge University Press, New York.
Wainer, H. & Spence, I. e. (2005b), *The Statistical Breviary; Shewing on a Principle entirely new, the resources of every state and kingdom in Europe; illustrated with Stained Copper-Plate Charts, representing the physical powers of each distinct nation with ease and perspicuity by William Playfair*, Cambridge University Press, New York.
Wang, P. C. C., ed. (1978), *Graphical Representation of Multivariate Data*, Academic Press, New York.
Wasserman, S. & Faust, K. (1994), *Social Network Analysis: Methods and Applications*, Cambridge University Press, Cambridge.
Wegman, E. (1990), Hyperdimensional Data Analysis Using Parallel Coordinates, *Journal of American Statistics Association* **85**, 664–675.
Wegman, E. J. (1991), The Grand Tour in k-Dimensions, Technical Report 68, Center for Computational Statistics, George Mason University.
Wegman, E. J., Poston, W. L. & Solka, J. L. (1998), Image Grand Tour, *in* Automatic Target Recognition VIII - Proceedings of SPIE, 3371, SPIE, Bellingham, WA, pp. 286–294. Republished, Vol 6: Automatic Target Recognition. The CD-ROM, (F. Sadjadi, ed), SPIE: Bellingham, WA, 1999.

Won, X., Fuhrman, S., Michaels, G. S., Carr, D. B., Smith, S., Barker, J. L. & Somogyi, R. (1998), Large-scale temporal gene expression mapping of central nervous system development, *in* Proceedings of the National Academy of Science, Vol. 95, pp. 334–339.

Wickham, H. (2006a), classifly: Classify and Explore a Data Set, http://www.R-project.org.

Wickham, H. (2006b), DescribeDisplay: R Interface to DescribeDisplay (GGobi Plugin), http://www.R-project.org/.

Wickham, H. (2006c), ggplot: An Implementation of the Grammar of Graphics, http://www.R-project.org.

Wilhelm, A. F. X., Wegman, E. J. & Symanzik, J. (1999), Visual Clustering and Classification: The Oronsay Particle Size Data Set Revisited, *Computational Statistics: Special Issue on Interactive Graphical Data Analysis* **14**(1), 109–146.

Wilkinson, L. (1984), SYSTAT: The System for Statistics, SYSTAT, Inc., Evanston, IL.

Wilkinson, L. (2005), *The Grammar of Graphics*, Springer, New York.

Wills, G. (1999), Nicheworks – Interactive Visualization of Very Large Graphs, *Journal of Computational and Graphical Statistics* **8**(2), 190–212.

Young, F. W., Valero-Mora, P. M. & Friendly, M. (2006), *Visual Statistics: Seeing Data with Dynamic Interactive Graphics*, John Wiley & Sons, New York.

Index

Adjacent Transposition Graph *see* datasets
animation 61, 111, 136, 147
Arabidopsis Gene Expression *see* datasets
Australian Crabs *see* datasets
average shifted histogram (ASH) 21, 71, 143

barchart 21
  spine plot 21
  stacked 23
Bayes Information Criterion (BIC) 113
brushing 19, 34, 35, 37
  conditioning 37
  database query 36
  geometric sectioning 37
  linked 36, 55, 58, 61, 78, 91, 95, 127, 137, 147
  persistent 39, 104, 107, 108
  selection sequences 44
  transient 39, 111

canonical correlation analysis 34
classification
  discriminant space 78
  error 69
  examining boundaries 97
  misclassification table 69
  supervised 63
  unsupervised *see* cluster analysis
classification methods

linear discriminant analysis (LDA) 34, 65, 77
logistic regression 66
neural network 88
quadratic discriminant analysis (QDA) 66
random forest 69, 83
support vector machine (SVM) 92
support vector machines 34
tree 66, 81
classification strategy 99
cluster analysis 103
  algorithms 111
  cluster characterization 124
  confusion table 122
  dendrogram 111
  hierarchical 111
  intercluster distance (linkage) 111
  interpoint distance 103, 105, 107
  model-based 113
  nuisance variable 103
  self-organizing maps (SOM) 119
  spin and brush 104, 107
Cluster Challenge *see* datasets
confusion table *see* cluster analysis
cross-sectional studies 134
cross-validation 69

data 1
data analysis
  data preparation 10
  exploratory (EDA) 3
  initial 13
  problem statement 10

186   Index

teaching   14
wisdom   8, 14
data mining   66
data snooping   13
datasets   153
    Adjacent Transposition Graph   172
    Arabidopsis Gene Expression   168
    Australian Crabs   17, 113, 154
    Cluster Challenge   172
    Flea Beetles   67, 131, 156
    Florentine Families   140, 173
    Italian Olive Oils   67, 155
    Morse Code Confusion Rates   147, 173
    Music   122, 170
    PRIM7   108, 157
    Personal Social Network   175
    Primary Biliary Cirrhosis (PBC)   161
    Rat Gene Expression   166
    Spam   162
    Tips   5, 19, 153
    Tropical Atmosphere-Ocean Array (TAO)   50, 159
    Wages   135, 164
dendrogram   see cluster analysis
dot plot   20
    jittered   20
    stacked   20
    textured   20
double decker plot   44
dragging points   43
drawing   43
drill-down   7

EDA   see data analysis
ensemble method   69
ESTAT   44
exploratory data analysis   10, see data analysis

Flea Beetles   see datasets
Florentine Families   see datasets

GeoVista Studio   44
GGobi   V
ggvis   147
graph   43, 139
    edge   139
    layout   139
    node   139

hierarchical clustering   see cluster analysis
histogram   5
    bin width   5
human–computer interaction   3
hypothesis testing
    alternative hypothesis   129
    null hypothesis   129
    test statistic   131

identification   41
    persistent   41
    transient   41
inference   3, 13, 129
information   1
initial data analysis   see data analysis
interpoint distance   see cluster analysis
Italian Olive Oils   see datasets

jittering   20, 87, 124

knowledge   1

labeling   see identification
linear discriminant analysis (LDA)   see classification methods
linkage   see cluster analysis
linked brushing   see brushing
linking
    m-to-n 38
    categorical   38, 137
    one-to-one   38
    point-to-edge   39
longitudinal data   134

MANET   44
misclassification   see classification error
missing values   46
    ignorable   48
    imputation   47, 55
    missing at random (MAR)   48
    missing completely at random (MCAR)   48
    missing not at random (MNAR)   48
    multiple imputation   59
    non-ignorable   48
    shadow matrix   49

model-based clustering   *see* cluster analysis
modeling   11
Mondrian   44
Morse Code Confusion Rates   *see* datasets
mosaic plot   15, 23, 44
multidimensional scaling (MDS)   120, 145
multiple views   2, 36
Music   *see* datasets

neural network   *see* classification methods

Olive Oils   *see* datasets
outliers   111, 120, 122

paintbrush   35
painting   39
panel data   *see* longitudinal data
parallel coordinate plot   24, 34, 73, 106, 126
PBC   *see* datasets
Personal Social Network   *see* datasets
plot interactions
    brushing   *see* brushing
    dragging points   *see* dragging points
    drawing   *see* drawing
    identification   *see* identification
    labeling   *see* identification
    painting   *see* painting
    tour   *see* tour
    view scaling   *see* view scaling
plot types
    ASH   *see* average shifted histogram (ASH)
    barchart   *see* barchart
    dendrogram   *see* dendrogram
    dot plot   *see* dot plot
    histogram   *see* histogram
    mosaic plot   *see* mosaic plot
    parallel coordinate plot   *see* parallel coordinate plot
    scatterplot   *see* scatterplot
    scatterplot matrix   *see* scatterplot matrix
    spine plot   *see* barchart
plots

arrangement   34
bivariate   21
conditioning   7
interactive   3, 15
model   12
multivariate   24
of categorical variables   17
of real-valued variables   17
presentation   14
univariate   19
PRIM7   *see* datasets
principal component analysis   22, 31, 34, 108, 120
projection pursuit   30
    indexes   30, 108
        central mass   30, 108
        holes   30, 108
        LDA   31, 131
        PCA   30

R   V
R package
    `DescribeDisplay`   XIII
    `Hmisc`   59
    `classifly`   97
    `e1071`   94
    `ggplot`   XIII
    `iPlots`   44
    `libsvm`   94
    `nnet`   88
    `norm`   59
    `randomForest`   83
    `rggobi`   V, XIII
    `tuneR`   171
random forest   *see* classification methods
Rat Gene Expression   *see* datasets
regression   8, 11, 34

scatterplot   5, 21
    conditioning   7
scatterplot matrix   25, 34
self-organizing maps   *see* cluster analysis
shadow matrix   *see* missing values
social network   139
Spam   *see* datasets
sphering   31
spine plot   23

188   Index

supervised classification   *see* classification
SVM   *see* classification methods
support vector machine (SVM)   *see* classification methods
SVMLight   93

TAO   *see* datasets
time series   134
Tips   *see* datasets
tour   18, 26, 29, 107, 147
   grand   29, 72, 76–78, 80, 85, 87, 88, 91, 95, 97
   manual   29, 31, 76, 82, 85, 97
   projection pursuit guided   29, 30, 74, 76, 80, 131
tree   *see* classification methods
trellis plot   34

unsupervised classification   *see* cluster analysis

view scaling   41
visualization   2
   cartographic   2
   data   2, 3
   flow   2
   information   2
   scientific   2, 3
   surface   2
   uncertainty   3
   volume   2

Wages   *see* datasets

XmdvTool   44

 **springer.com**

Bayesian Computation with R

Jim Albert

Bayesian Computation with R introduces Bayesian modeling by the use of computation using the R language. The early chapters present the basic tenets of Bayesian thinking by use of familiar one and two-parameter inferential problems. Bayesian computational methods such as Laplace's method, rejection sampling, and the SIR algorithm are illustrated in the context of a random effects model. The construction and implementation of Markov Chain Monte Carlo (MCMC) methods is introduced. These simulation-based algorithms are implemented for a variety of Bayesian applications such as normal and binary response regression, hierarchical modeling, order-restricted inference, and robust modeling.

2007. 280 pp. (Use R!) Softcover ISBN 978-0-387-71384-7

Graphics of Large Datasets: Visualizing a Million

Antony Unwin, Martin Theus, and Heike Hofmann

This book shows how to look at ways of visualizing large datasets, whether large in numbers of cases or large in numbers of variables or large in both. Data visualization is useful for data cleaning, exploring data, identifying trends and clusters, spotting local patterns, evaluating modeling output, and presenting results. It is essential for exploratory data analysis and data mining. Data analysts, statisticians, computer scientists-indeed anyone who has to explore a large dataset of their own-should benefit from reading this book.

2006. 271 pp. (Statistics and Computing) Hardcover
ISBN 978-0-387-32906-2

Analysis of Phylogenetics and Evolution with R

Emmanuel Paradis

The book starts with a presentation of different R packages and gives a short introduction to R for phylogeneticists unfamiliar with this language. The basic phylogenetic topics are covered: manipulation of phylogenetic data, phylogeny estimation, tree drawing, phylogenetic comparative methods, and estimation of ancestral characters.

2006. 211 pp. (Use R!) Softcover ISBN 978-0-387-32914-7

Easy Ways to Order▶ Call: Toll-Free 1-800-SPRINGER • E-mail: orders-ny@springer.com • Write: Springer, Dept. S8113, PO Box 2485, Secaucus, NJ 07096-2485 • Visit: Your local scientific bookstore or urge your librarian to order.